OXFORD BIOLOGY PRIMERS

Discover more in the series at
www.oxfordtextbooks.co.uk/obp

Published in partnership with the Royal Society of Biology

PRESENTING SCIENTIFIC DATA IN R

OXFORD BIOLOGY PRIMERS

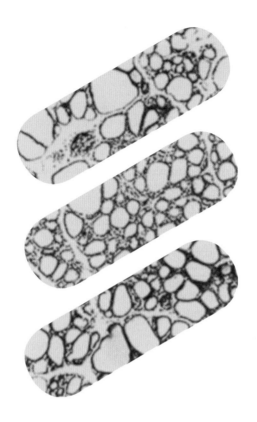

PRESENTING SCIENTIFIC DATA IN R

Rosalind K. Humphreys and Graeme D. Ruxton

OXFORD
UNIVERSITY PRESS

Royal Society of
Biology

OXFORD
UNIVERSITY PRESS

Great Clarendon Street, Oxford, OX2 6DP,
United Kingdom

Oxford University Press is a department of the University of Oxford.
It furthers the University's objective of excellence in research, scholarship,
and education by publishing worldwide. Oxford is a registered trade mark of
Oxford University Press in the UK and in certain other countries

Impression: 1

Published in the United States of America by Oxford University Press
198 Madison Avenue, New York, NY 10016, United States of America

British Library Cataloguing in Publication Data
Data available

Library of Congress Control Number: 2022931699

ISBN 978-0-19-887047-0

Printed in the UK by
Bell & Bain Ltd., Glasgow

Figures, screenshots, and image captures were created using R: R Core Team (2020)
R: A Language and Environment for Statistical Computing. R Foundation for Statistical
Computing, Vienna, Austria: https://www.R-project.org/.

To Mum, Dad, and Andrew; and to Katherine

PREFACE

Why read a book on drawing graphs

Vision is a very important sense to humans, and we are highly adept at absorbing information presented visually. For that reason, when presenting data that can be used to explore a scientific question it is very common to employ graphical representations of that data. It is rare to find a scientific publication that does not feature figures displaying aspects of the data. However, we biologists often do not have any formal training in presenting these figures as effectively as possible. This is very much at odds with statistical exploration of data, where most students at all stages will undertake entire modules devoted to statistics, and many later-career researchers will invest time in expanding their statistical skills. In contrast, it seems there is a presumption that presenting data in effective figures is simply something that comes naturally, and perhaps improves with practice. We disagree. We think that scientists at all stages can improve the communication of their science through taking more care over their figures. We also think that they should feel strongly motivated to take more care. Better-designed figures should lead to higher scores in lab reports for junior undergraduates, and easier passage through peer review for later-career scientists authoring their own grant proposals or scientific papers. Taking more care with your figures should:

- Make your work more eye-catching and allow it to stand out from the crowd.
- Enhance the ease with which readers understand and remember your work.
- Enhance your own understanding of your data, and offer more insight into the scientific questions you are interested in.

Why you should read this book specifically

We are all time-stressed—in light of this, we have made this book as compact as we can. However, compact should not mean trivial, and we are confident that no matter how many graphs you have drawn in the past, there should be material in this book that will help you think more critically about the graphs that you draw in the future—thereby improving all the graphs you draw from now on. We have made the book compact by covering what we consider are fundamentals, without trying to be exhaustive. As part of this, we do not get tangled up in statistical detail; our focus is effective data presentation as there are plenty of resources already dedicated to statistics. Our graphics package of choice (R: more on why we picked this in chapter 1) is extraordinarily flexible—we do not attempt to cover every possible customization option. Rather we focus on raising awareness of - and technical skills in - what we consider are core issues, in the belief that your raised confidence and abilities will help you to find answers to any issues that we don't cover (through a little exploration in your favourite search engine). We also provide extensive supplementary online resources at www.oup.com/he/humphreys-obp1e that we think will cement concepts or stretch understanding for some readers.

If you flick through this book you will see plenty of R code, and we do see this book as providing a technical function in helping you see how you communicate with R to achieve desired aspects of your graph. However, we do not see this book as only, or even primarily, a technical 'how-to' manual. We think the biggest challenge isn't in getting R to produce the graph you want, but in knowing what will make a really effective graphical representation of your data for a specific purpose. Thus, this book is very much about discussion of what makes an effective graphical representation of scientific data—and this has nothing to do with the technicalities of how to achieve this in R or any other computer package. Even if you do not want to use R, this book should help you design much better data figures in future, as well as improve your ability to critique and evaluate other scientists' graphs. As an added bonus, we offer exactly the same benefits for tables of scientific data (see chapter 2).

How to use this book

As we have said, we know you are time-stressed, but we have organized this book (and kept it compact) so that we think you will gain benefits from working through the book once from start to finish. We hope that even the discussion of graphic types that you don't ordinarily use should offer value to you in enriching how carefully you think about your own graphing, and how effectively you interpret other scientists' graphs. However, we have also thought hard about the organization of the material, the titles we give different sections, and cross-referencing to make this a useful resource when you just need to quickly solve a particular issue you have with the graph you are currently working on.

In terms of organization, you will see that we devote most chapters to a commonly used type of graph (or graphs). We think the easiest way to quickly decide on the most suitable type of graph in any situation is to be clear about the type of data that you have—and we present our thinking about that early in chapter 1. The unusual chapters in this regard are chapters 7 and 8—both of these tackle aspects of graph design that span many different graph types. We think the material in chapter 7 will be useful to the vast majority of readers on a regular basis, whereas the material in chapter 8 might be very useful to you sometimes, but not necessarily in every graph that you draw. We include core material in the book, but provide online resources where we think some readers will appreciate delving deeper or testing their understanding with further examples. Also, all of the R script files we provide in the online resources are extensively annotated, so that you can make clear links between the code and its function outside of the context of the main book (where key features of code are explained in the main text).

Lastly, we hope you use this book with a smile. We do not want to give the impression that graphing is much harder than you thought it was, or that there is a whole gamut of pitfalls that you will blunder into. Rather, we want to foster the idea that producing figures to support your scientific investigations should not be a dull and routine chore; it should be intellectually stimulating, rewarding, and fun.

ACKNOWLEDGEMENTS

This book was largely inspired by the response to a day-long course we ran at the Scottish Schools Education Research Centre (SSERC) in February 2019. The course was delivered to an audience of high-school science teachers from across Scotland and covered how to effectively visualize scientific data in R. The enthusiasm of, and positive feedback from, the teachers confirmed to us not only that taking greater care with the production of figures can be intellectually rewarding and fun, but also that doing so using R is an incredibly effective way to achieve publication-quality figures (even for total R-newbies!). As such, we first of all need to thank everyone who attended that course, and Paul Beaumont and Jim Stafford from the SSERC for organizing it.

From the very beginning of writing this book, we have received excellent support and advice from several members of editorial staff at Oxford University Press. Lucy Wells and Sophie Ladden were great guides to us in the initial development stages, helping to get the book off the ground and focus our ideas. Giulia Lipparini subsequently provided insightful recommendations and valuable discussion, which considerably improved the content and organization of the chapters. Lucy Wells then took the helm again, contributing carefully considered suggestions and steering us through the final stages of production. Her patience, wisdom, and understanding allowed us to realize our ambitions with this book. The OUP team also secured four fantastic reviewers, whose enthusiastic engagement with the material and perceptive comments stimulated significant and hugely beneficial changes in the book. We are also thankful to the design, production, and copy-editing staff at OUP.

We would also like to extend our profound thanks to Rhys Hague, for the considerable time and thought they put into providing us with valuable feedback on all of the chapters and supplementary material. Their conscientious input and enthusiasm was a great aid in finalizing all of the material.

Any errors or omissions in the book or its supplementary material are our own.

TABLE OF CONTENTS

INTRODUCTION AND GETTING STARTED

1

Learning objectives

By the end of this chapter you should be able to:

- Explain what makes a good figure and figure **caption**.
- How to choose the right data visualization tool for the data you have.
- Appreciate why R is our tool of choice for good graphing.
- Choose colours effectively for your type of data.
- Appreciate the 'simple, but not too simple' rule of thumb when designing figures.

1.1   Introduction: the importance of good graphing

Data is the heart of all science. It is what we gather, what we draw findings from, and ultimately what we use as a springboard from which to grow our understanding. But the process of collecting data is often messy and complex, generating a great mass of data that we then must analyse and consider as a whole in order to understand what we have found. This is where effective data visualization is an invaluable tool. Not only do figures provide an excellent way to organize data but, more importantly, such visual aids can be really useful when it comes to interpreting data and looking for patterns. You will never find a data set of raw values that would not be easier to interpret as a graph or well-presented table. Data visualization enables us to distil our data into a clear summary that communicates our findings in a more accessible and engaging way than lists of undigested values alone ever could.

For any data set, there are often multiple presentation options that you might want to consider. The best option to take will often depend both on the type of data you have collected (see section 1.3 for more discussion of data types) and what components of that data you want to focus on. Throughout this book, where we describe multiple presentation options for a given type of data, we also discuss the relative advantages and disadvantages to each option, to help guide you in making well-informed data visualization decisions. However, there are some general tips you should be sure to follow when creating any figure, which we discuss in Scientific Approach 1.1. The chapters of this book

Scientific Approach 1.1
General approaches to good use of figures

We think there are a number of general points that you can bear in mind, no matter the type of figure you are creating, to improve your use of figures and the quality of your figures:

- Number each figure in your document, and make sure they are referred to sequentially in the document.
- Make sure you give each figure a full and informative title (or caption) that gives a summary of its content without reference to the main text of your report.
- Make sure both axes are sensibly proportioned and labelled.
- Make sure units and sample sizes are given (where appropriate).
- Add **grid lines** (sometimes called 'graph lines') if this will significantly improve interpretation of measurements.

- Make sure any figure is accompanied by an appropriate caption (see section 1.2 for an explanation of our use of the term 'caption').
- Make sure there is a **legend** (aka a key) if this is needed to interpret the figure.
- Be consistent in nomenclature, organization, colours, and symbols between figures showing similar data within the same report.
- Always check figures and associated captions carefully for typos.
- When referring to figures in the text, don't be too brief—don't say just 'see Figure 1'. Guide the reader more by saying, for example, 'Notice in Figure 1 that the green bars for female cinema attendance are substantially higher than the blue bars for male cinema attendance only in the age categories of 45–54 and 55+.'

provide walk-through steps for how to create both simple and more refined versions of the most commonly used figure types that are effective and clear. We also provide extensive supplementary online resources at www.oup.com/he/humphreys-obp1e that you can use to develop your understanding and skills—we will direct you to these at relevant points throughout the chapters.

 Key point

You want a graph that provides a clear summary of your data, and that communicates interesting features of that data in an accessible and engaging way.

1.2   Good figures have good captions

All the figures presented in this book are accompanied by figure captions. Captions are essential accompaniments to any figure that you produce, concisely providing readers with enough information to understand the figure without recourse to the main text of your report; we provide some general tips on producing effective figure captions in Scientific Approach 1.2. The form of a figure caption can vary between circumstances. For example, in a report or scientific paper when you are presenting data the caption should be fully detailed and written in a neutral style—some of the figure captions in this book are of this form. However, in this book our figures are often highlighting issues of presentation rather than the data itself, and are seen as part of a conversation with you, so we frequently adopt a shorter and more informal style. Just remember to keep your own captions formal and professional unless your context definitely suggests a different approach.

Figure captions are also sometimes referred to as figure legends. However, in R (the statistical program used throughout this book) the word 'legend' refers to a chart's key—that is, a feature of some figures that explains what different colours/points/line types represent. To avoid confusion, text descriptions that accompany figures will be referred to as 'figure captions' throughout this book, and 'figure legends' will refer to a feature providing additional information to allow interpretation of the data. For consistency, we also refer to the concise descriptive text accompanying tables as 'table captions' in section 2.3.1.3; the suggestions set out in Scientific Approach 1.2 apply to table captions too.

Scientific Approach 1.2
General approaches to effective captioning

All figures should have a caption. We think there are a number of general points that you can bear in mind, no matter the type of figure, to improve the effectiveness of your captions:

- Provide enough detail to support your figure independent of the main text of your report. That is, the figure should be intelligible in its entirety without a reader having to refer to the main text. Hopefully examination of some of the captions written in a more full and formal style in this book should show you what we mean (in particular, we write captions in this way for all of the 'Authors' attempts' documents that accompany chapters 2 to 6 in the online resources).

- At the same time, do not repeat large sections of the main text—your caption should (as concisely as possible) provide the minimum amount of information required to explain what the figure shows in the context of your work.

- Ensure that any use of symbols, colours, lines, patterns, error bars, or any other potentially ambiguous features in the figure are clearly explained in your caption.

- Ensure that any abbreviations, nomenclature, and units in your caption are consistent with the main text.

- Always check figure captions carefully for typos.

- Captions should be placed above a table, but below a figure (charts, graphs, images, etc.).

- Remember to check that you have permission for any components of a figure that have been reproduced from someone else's work and reference the original source in your caption.

Key point

A figure always needs a caption, and we encourage you to take the time to write an effective one.

1.3   Different types of data: what do I have and what are my options?

Not all figure types are suitable for all data sets, and deciding which presentation tool suits your data best is not always immediately obvious. However, there are some fundamental questions you can ask yourself about your data to help you work out what your options are. Figure 1.1 is a decision tree you can use to help you decide what data visualization tools, and thus what chapters of this book, might be most useful to you based on the data you intend to present. First of all, you need to ask yourself what sort of data you have. We briefly discuss the different data types below, but if you want to solidify your ability to identify data types then check out the guide 'Data types' in the online resources, which contains a fuller discussion with more detailed examples.

1.3.1 Quantitative data

If your data is quantitative, this means it involves measurements on a numerical scale. These measurements can be:

- Continuous: That is, measurements recorded could theoretically have an infinite number of possible values within a range, depending on how fine-scaled your measurement equipment is. The units in which your measurements were taken could logically be subdivided and recorded as decimals or fractions. Examples of continuous data include temperature, height, and weight.

- Discrete: This is where measurements in data include only integers and the scale on which measurements were recorded constitutes a finite number of values that cannot be meaningfully subdivided. A good example of this might be number of children—you cannot have only 0.8 of a child in your household!

1.3.2 Qualitative data

If your data is qualitative, this means it involves counts arranged into categories (and is therefore sometimes referred to as 'categorical' data). Sometimes these categories are assigned numerical values as their names, but that is not to imply that they sit on a logical and equally spaced numerical scale with consistent units (as quantitative data does). If we consider the number of children in different households, this data is quantitative rather than qualitative, because you can think logically about the exact distances between possible values. The difference between 1 and 2 children per household is the same as the difference between 2 and 3; the unit (1 child) is consistent. However, dress sizes are an example of qualitative data, because the difference (e.g. in material used or waist size) between dresses of sizes 10 and 12 isn't the same as the difference for sizes 12 and 14. We have separate categories of dress size, but differences between these are undefined. Qualitative data can be:

- Ordinal: The categories to which counts or observations are assigned have a logical rank ordering to them. An example of this would be shoe size or dress size.

- Nominal: This is where the different categories have no logical rank order to them. Examples of nominal data include blood group, eye colour, and ethnicity.

The decision tree in Figure 1.1 should help point you in the right general direction, but further details on the relative advantages and disadvantages of different figure types are provided in their respective chapters to help you make the right decision for your data. Further, if you are unsure which variable should

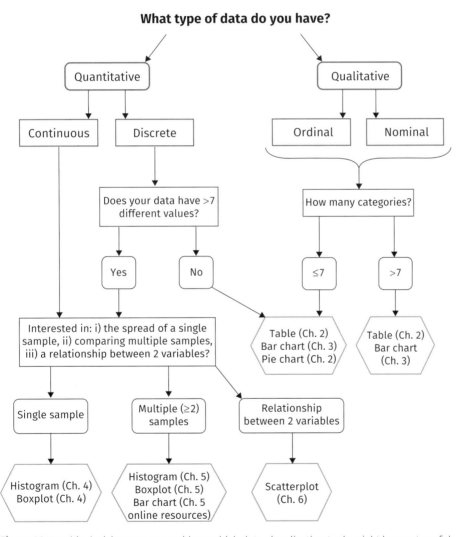

What type of data do you have?

Figure 1.1 Use this decision tree as a guide to which data visualization tools might be most useful for the type of data you want to present (and which chapter(s) in this book and/or online resources can walk you through them). Starting at the top, you'll need to think about what type of data you have and, from there, ask yourself the subsequent questions to lead you to an orange hexagon with possible options.

be presented on which axis of a figure, Bigger Picture 6.2 in chapter 6 can give you some useful pointers.

Key point

Thinking both about the type of data you have, and the features of that data you want to communicate, can take you a long way towards finding a good design for your figure.

1.4   Why R is a good data presentation tool

As discussed in section 1.1, graphing is a really important skill in the sciences. Graphs enable both effective organization and interpretation of data. One way to get really professional-looking graphs relatively easily is through using R. As well as being the most comprehensive and reliable statistics package in the known universe, it is also the most flexible graph-drawing package. Another bonus is that it is completely free to anyone, and works on both PC and Mac, regardless of operating system. You can also save the code you use to produce any graph and come back to edit or reproduce it whenever you like. If you take R code that you wrote three years ago on a PC and email it to your friend today who runs an updated version of R on their Mac, then it will still work seamlessly for them.

RStudio is a user-friendly interface that we highly recommend, as it makes interacting with R much more straightforward. RStudio also standardizes the way things work across different computer platforms. This means that the descriptions we provide in this book for how to do something in RStudio will work for you, whether you are using a PC, MAC, or LINUX system. For these reasons, we will use RStudio to interact with R throughout this book. Before you can install RStudio, however, your system needs to have an up-to-date version of base R installed. This is very easy to do, whether you have used R previously or not—see the 'R Basics' guide in the online resources for this chapter for instructions on downloading and installing both R and RStudio.

Throughout this book, mention of R will be in reference to the overall statistical package that is the base of RStudio and which processes the code we input, but RStudio will be specified when discussing details or steps specific to this interface's layout and functioning. All figures provided in the book and online resources are reproducible in RStudio with the code and steps provided.

We prefer to draw graphs in R using what is called base graphics—these are the graphical functions supplied in the core module of R. Increasingly over the last decade, many R users have adopted the graphical functions in an add-on package called 'ggplot2' (Wickham 2016). The ggplot2 package is part of a group of add-on packages that together embody the so-called tidyverse philosophy of handling data. We do not claim that our approach produces intrinsically better graphs than can be produced in ggplot2. You can produce essentially identical-looking figures by both methods—the question is which approach is easiest. Unquestionably, ggplot2 will be more attractive to those who have bought into the tidyverse approach more generally. It may also be attractive to some researchers who use R as part of their daily routine. However, for the less hardcore R user, we believe that the base graphics approach that we adopt is easier to grasp and remember.

Key point

R is not just a great package for statistical analysis—it lets you create fabulous figures too.

1.5   Assumed prior experience of R

In writing this book, we have made some assumptions about readers' familiarity with the interface of RStudio and prior experience with R. Specifically, we assume that you are comfortable with the following:

- Downloading and installing R and RStudio.
- Operating RStudio, entering simple data, and performing simple calculations using R.
- Using R scripts to edit and save commands.
- Producing and saving graphics in RStudio.
- Loading data into RStudio.
- Installing and activating packages in RStudio.

However, do not despair if our assumptions are incorrect for you and/or if you are completely new to R. We provide a couple of documents in the online resources which will get you entirely up to speed (or serve as handy refreshers if you have used R little or infrequently). The first is called 'R Basics', and this (alongside its associated R script) covers all of the bullet points listed above. The second is a 'Walk-through example', which will let you put into practice some of the 'R Basics' material and get you thinking about some of the more general concepts related to figure production covered in this chapter. This example uses data that we provide from a real experiment (seed_data.csv) and also comes with an R script file. Both R scripts are annotated with detailed comments to explain the code we use, and you will find that this is true of all the R script files we provide in the online resources to accompany each chapter.

Whether you have used R many times before or not, we wish to draw your attention to a line of code that we use commonly throughout the book when loading in data sets. There are a number of ways you can load your data into RStudio, including very specific lines of code that will establish a working directory from which to extract files from your computer (see Further Reading in the 'R Basics' document from the online resources if you are interested in these).

However, to us the simplest method is to run the following line of code, which can be used wherever your data file is saved on your computer:

```
alldata <- read.table(file.choose(), header = T, sep = ",")
```

This is not as spooky as it looks. 'alldata' is just a name that we want to use for our collection of data—we could have used any name. Next, the two symbols '<-' essentially form an arrow (or 'assignment character') that points to the name that we want to assign to the following data. 'read.table' is a clever function for reading data into R. 'file.choose()' tells R to open a box and let you choose the file you want to open in RStudio, wherever it is saved. 'header = T' tells R to use the names you have given to the columns in the data file to identify different parts of the data, and 'sep = ","' warns it to expect that numbers will be separated by commas (rather than say TABs) in the file.

We will use **read.table** and **file.choose()** when loading in the data sets we provide throughout the book, but it would also work for whatever data set you want to load into R. So long as your data is arranged appropriately in an Excel spreadsheet, and saved as a .csv file, this code will work (see the 'R Basics' document in the online resources for more detailed advice on preparing data for use in R).

> **Key point**
>
> We assume that you have a little familiarity with R already, but if you don't we offer a quick start-up guide in our companion material in the online resources.

1.6   Choosing colours effectively

Often you are going to want to control the colours used in your figures. R has a massive range of 657 colours available—check out this full list of colours at http://www.stat.columbia.edu/~tzheng/files/Rcolor.pdf.

We also provide a one-page guide listing some colours available in R that we use again and again (including all the colours used in this book and its online materials) as an online resource called 'Colour Guide'.

Colour can add so much to the visual appeal of figures; it can also be an effective way to encode information, and increase clarity. We can offer some tips to stimulate you to use colour well.

An initial consideration is whether or not using different colours will be informative to viewers. If colour will not in itself refer to some additional variable that is not otherwise shown, then you should not confuse your figure by using multiple colours (unless they are required as a component of your figure, such as a pie chart or grouped bar chart). If colour can be used to add valuable information, you should have a think about what type of data you have on your additional variable (see section 1.3 and the 'Data types' document in the online resources for details on different types of data). Whether your data has a logical order to it or not can be a useful guide as to whether it would be meaningful for your colours to be, respectively, similar in shade and/or hue or distinct from one another. That is, the data type will determine whether your figure should most naturally be coloured using a *qualitative, sequential,* or *diverging* palette.

Let's demystify these terms. Figure 1.2 shows three scatterplots, each constructed from some imaginary data concerning the distance volunteer biologists managed to run in 15 minutes at different environmental temperatures. In scatterplot A, an additional variable, the area of biology that different volunteers specialized in, is coloured with a qualitative colour palette. Although the temperature and distance measurements are certainly quantitative, the volunteers' study area is a nominal qualitative variable (see section 1.3 for more on data types) and therefore there is no reason to choose colours that are clearly related in shade or hue. For that reason, distinct colours have been chosen to identify the volunteers' subject areas. We chose a group of colours that we felt were easy to tell apart and assigned them to the subject areas in no particular order. Scatterplot A suggests that subject area does not seem to be related to the distances run—the trend of shorter distances at higher temperatures generally applies across all the data.

In scatterplot B, an alternative additional variable of caffeine consumed in advance of the run is coloured with a sequential colour palette. In this hypothetical study, volunteers were assigned to five different treatment groups, that were each asked to consume different quantities of caffeine before their 15 minutes of running time. In this case, the additional variable in the study (caffeine consumed) has a logical order to the different levels—that is, the caffeine level consumed by the different treatment groups increased from one group to the next. For this reason, a sequential gradation of colours has been used to identify the different treatment groups, starting with lighter hues for the lower-caffeine groups and increasing in darkness as the caffeine quantity increases. We think that the logical ordering to the hues used helps you pick out the general trend in the effect of caffeine easily. From scatterplot B, it looks as though increased caffeine consumption may increase the distance volunteers ran at any given air temperature.

In scatterplot C, a third alternative additional treatment variable spanning both caffeine consumption and prior exercise at various levels is coloured with a diverging colour palette. In this version of the imagined study, volunteers were again assigned to five different treatment groups. This time, though, two groups were asked to consume drinks high in caffeine (energy drinks and coffee), two groups were asked to exercise to different degrees (mild and heavy), and one group did not exercise or consume caffeine (control). Here, the control group serves as a logical central measure between the groups experiencing either caffeine consumption or presumed fatigue, so it makes sense that the opposite sides of this reference value should be coloured with two distinctive hues to reflect these two quite different deviations from the control condition. Each hue fades in shade towards the central control and darkens away from it as the broader group type 'caffeine consumption' or 'presumed fatigue' becomes more extreme and different from the control condition. In scatterplot C we can see that exercising in advance of the run appears to have reduced distances volunteers ran compared to the control group, while consuming caffeine appears to have generally increased the distances run by volunteers.

In Scientific Approach 1.3 we detail further when these different palettes can usefully be applied, as well as offering some tips specific to each palette type, the general rules you can consider when choosing colours, and ways in which you can improve the accessibility of your figures by attending to colour blindness. There is no 'perfect' colour scheme that will suit every context and every viewer, but we can—and should—try to use colour informatively and thoughtfully when designing figures. One of the great bonuses of using R to produce figures is that if we are ever in a situation where a problem arises with the colours we have chosen, we can easily fix the problem by simply editing the colours in the original R script file and rerunning the code. Although there are fewer established rules of thumb, colour choice is also worth thinking about when you are presenting data in table form, which we also touch on in Scientific Approach 1.3.

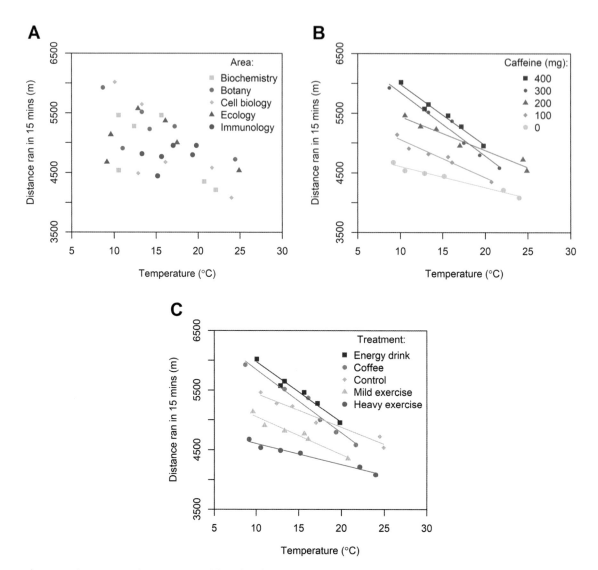

Figure 1.2 Three scatterplots constructed from imaginary data collected on the distance volunteer biologists managed to run in 15 minutes at different environmental temperatures. The additional variable measured differs for each hypothetical situation. **A: Qualitative colour palette**. In this version of the study, the area of biology the volunteer runners work in was also recorded. **B: Sequential colour palette.** In this version of the study, volunteers were assigned to five different treatment groups, that were each asked to consume different quantities of caffeine before their running time. **C: Diverging colour palette.** In this version of the study, volunteers were assigned to five different treatment groups: two groups consumed drinks high in caffeine before running (energy drinks and coffee), two exercised to different degrees before running (mild and heavy), and one group did not exercise or consume caffeine (control).

Scientific Approach 1.3
General concepts in colour choice

Use of colour is an important, creative and satisfying aspect of data visualization, and it is useful to consider the type of data that you are presenting in order to make effective colour choices. Some rules of thumb for different colour palettes are outlined below.

Qualitative palettes (a group of colours without a logical ordering):

- Good for nominal categorical data.
- Choose distinct colours.
- As well as selecting different hues of colour, additional variation can be achieved through choosing lighter or darker shades of a colour. However, try not to vary the lightness of hues too much with nominal data, as this may suggest that some colours are more important or of greater value than others. Also, avoid having two colours with the same hue but different lightness where your data is nominal unless the values associated with the two colours are related in some way.

Sequential palettes (a group of colours with a single logical ordering):

- Good for ordinal or quantitative (numeric) data.
- Choose colours on a continuum of lightness/shade, hue, or both.
- It is most intuitive to use light colours for lower values and dark colours for higher values.

Diverging palettes (two or more logical orderings of colour, each beginning from the same baseline colour):

- Good for quantitative data where there is a meaningful central, or baseline, value, either side of which your data is spread.
- Choose two distinctive hues, one for each side of the central value.
- Essentially, a diverging palette is made up of two sequential palettes with a shared end point at the central value.

- The central value will usually be a lighter colour, so that gradually darker colours either side indicate a greater distance from that centre. Ideally make the centre something like light grey rather than white, so that it still clearly represents values against the graph's (normally) white background.

General guidelines for choosing colours

What constitutes the perfect colour scheme for any given situation is highly subjective, and colour choices that improve the accessibility of a figure to one viewer may hamper interpretation by another. What is most important when choosing colours is that you avoid some common pitfalls and make sure that your figures are otherwise well designed to effectively communicate their data. Here are some general guidelines to stick to:

- Use different colours only when they correspond to differences of meaning in the data.
- Try to limit your maximum colour palette size to seven—when more colours than this are needed, it becomes harder to distinguish between colours and read a graph quickly. If you need more colours, consider whether your data would be better presented in table form.
- Be consistent with your use of colour across all graphs showing the same variables within a document.
- Use intuitive colours that readers may associate with your data anyway, e.g. environmental data = green, water-related data = blue, red = bad, green = good. However, avoid potentially offensive or stereotypical choices, such as the pink–blue combination to colour-code gender (see Further Reading for a link to an interesting blog post about this).
- If adding text, make sure its contrast to the background is high to make it readable.
- Wherever possible, attend to colour blindness to improve your plot's accessibility.

Attending to colour blindness

Around 4 per cent of people have some form of colour blindness. To make your graph accessible to colour-blind people, consider these tips:

- Vary a dimension other than hue alone (such as lightness/shade) to distinguish between values.
- Avoid using a combination of red and green in the same graph.
- Consider using the R package '**viridis**' (Garnier et al. 2021), which has a range of colour palettes that are easily readable to colour-blind people.

General rules on choosing colours for tables

- Use a background colour that contrasts highly with your text, so that it can be easily read.
- Make sure your non-data text is clear, but not so eye-catching that it distracts from the data.

 Key point

Using colour can add more information to a figure, make features clearer and easier for the reader to absorb, and/or improve the visual appeal.

1.7   Chart design and chart junk

1.7.1 Aim for your figures to be simple, but not too simple

When designing any figure, the primary goal is to present your data in a well-constructed way that communicates the important aspects of the data clearly to your readers. Lots of people will tell you that the best way to do this is to be as minimalist as possible in terms of the ink on the page. Many papers rail against redundant decoration of figures (often referred to as 'chart junk'), arguing that any details that do not tell viewers anything new about the data should be excluded. However, while we agree that any redundant symbols or extraneous graphical elements should be avoided, we advise that figures should follow a 'simple but not too simple' rule of thumb. Minimalism should play some role in your design choices, but this should not lead to you producing figures that are ugly or unpleasant to try and interpret.

As an example, Figure 1.3 shows three bar charts displaying the same information. Bar chart A includes several design embellishments that either contribute nothing valuable to a viewer's interpretation of the data (i.e. arrows on top of bars, country flag images, background image, bold and colourful frames around the plotting area and entire figure respectively) or would be better positioned in the figure caption (e.g. the text explanation of GDP). To our eyes, bar chart A is certainly 'too much'. By contrast, bar chart B is a minimalist design, stripped of all features not absolutely essential to conveying the measured data values. It lacks useful details (such as axis labels and a reference line at zero) which, although they do not depict the measured data values themselves, are important components for viewers interpreting the relationship between the presented values.

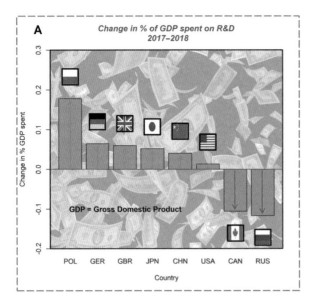
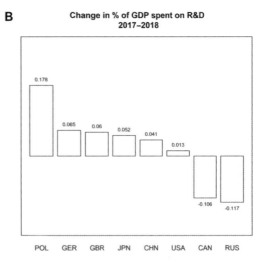
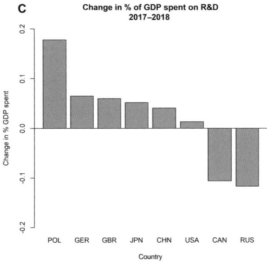

Figure 1.3 Three bar charts all showing the same data on the change in the percentage of GDP (gross domestic product) a selection of different countries spent on R&D (research and development) between 2017 and 2018 (OECD 2020). **A:** Bar chart A contains several unnecessary embellishments and is too cluttered. **B:** Bar chart B is minimalist, but also excludes useful features and is visually stark. **C:** Bar chart C follows our 'simple but not too simple' rule.

A figure should be designed to make any trend between values clear; if the intention is simply to present the raw data values (as bar chart B seems to) then these would be better placed in a table (see section 2.3 for how to produce effective tables). Bar chart B is also visually stark; by removing the fill-in colours of bars and enclosing the chart in a box of equal salience to the bar outlines, the minimalist design has made the bars unattractive to look at.

Turning to bar chart C, here we have a chart that was produced following our 'simple but not too simple' rule of thumb. It includes valuable details that are neither useless nor damaging (such as axis labels and a reference line at zero) and visually appealing features (such as a coloured fill-in of bars) without being too cluttered by unnecessary embellishments (like bar chart A) or too stark (like bar chart B). In fact, the details it includes that are not present in bar chart B are valuable, both to the interpretation of the figure and its visual appeal. Given that space did not need to be left to include features such as images or values as text, we were also able to draw bar chart C against a shorter y-axis range—this allows the bars to fill up more of the plotting area, enabling easier comparison of the relative percentage changes represented by bars both above and below the reference baseline.

Some of the key components of bar charts B and C were inspired by Stephen Few's examination of 'chart junk' (Few 2011). Note also that, while these figures are presented with figure titles, most figures included in a document or scientific journal should only be accompanied by a figure caption (see section 1.2 and Scientific Approach 1.2 for our advice on writing these). Titles should usually be reserved for situations where readers need to quickly interpret what a figure shows, possibly from a distance (e.g. during a presentation), but here we use them to contrast unnecessary embellishment in bar chart A with the formal style of bar charts B and C.

When producing figures, therefore, aim for a 'simple but not too simple' design. Figures you produce should:

- Not prioritize minimalism over visual appeal and clear interpretation.
- Always have clearly labelled axes.
- Include reference lines for significant values where necessary (e.g. zero, where there is a clear division between positive and negative values).
- Never replace axes with border boxes that have the same line thickness and colour as those in the figure itself.

But your figures also should:

- Not include unnecessary components that do not improve a viewer's ability to interpret the data.

1.7.2 A warning against cosmetic use of 3D

One particular embellishment that we consider unnecessary is 3D effects on figures. You will have seen plenty of bar charts where each rectangular bar has become what looks like a solid cuboid. Many programs now include design tools that allow the production of 3D graphs, including pie charts, bar charts, scatter graphs, and more. However, this is only very rarely a useful option for figure design. While 3D appears a superficially 'snazzy' and sophisticated-looking effect, it actually warps the position of data against the axes, making it very difficult to read values accurately. For this reason, we warn against its use in any context, particularly because there are a great many other (less misleading) ways to make your figures look visually appealing—as we hope to demonstrate in the later chapters of this book! The only context in which 3D is useful is when it is a component necessary to the display of data, such as in displays of terrain or 3D structures—here 3D is not just a flashy presentation choice, it is needed for appropriate visualization of the data and is essential to the figure production. We do not cover such specialist data types in this book, as they could fill a book in and of themselves!

Key point

Aim for figures that are neither stark and lifeless nor cluttered with needless distractions.

 Chapter Summary

- Data visualization enables us to organize data into a clear summary that communicates our findings in an accessible and engaging way.
- An informative caption is an essential accompaniment to any figure.
- The best type of figure to use depends on what type of data you have and what you want to communicate.
- R is a hugely flexible graph-drawing package and can be used alongside the user-friendly interface RStudio to produce professional-quality figures.
- Choosing colours carefully can allow you to clearly communicate extra information, increase clarity, and increase the visual appeal of figures.
- We recommend that your figures follow a 'simple but not too simple' rule of thumb—chart junk should be minimized, but not to the extent that figures are dull or unpleasant to try and interpret.

 Online Resources

The following online resources are available for this chapter at www.oup.com/he/humphreys-obp1e:

- Data types
- R Basics
- R script for R Basics
- Walk-through example
- seed_data.csv
- R script for walk-through
- Colour Guide
- R script for Figs (and associated .csv, .jpg, and .png files)

 Further Reading

Walk-through example

If you are keen to cement (or refresh yourself of) some of the basics of using R and/or the concepts outlined above, check out the online resources for this chapter. We provide a document called 'Walk-through example' alongside an R script and data set (seed_data.csv) for you to work through. See also Bigger Picture 1.1 for some general hints and tips on using R.

Bigger Picture 1.1
Some general R hints and tips

There are a few handy tricks that we find make our use of R easier, but that apply to all our use of R, not just creating graphs:

- Save your R code in an (extensively commented) script file. The code included in the chapters of this book is not annotated, as we explain key features in the main text, but all of the R scripts we provide online are carefully annotated using the # symbol (this is further explained in the 'R Basics' guide in the online resources).

- When you need to choose names for things, choose informative names.

- Make sure your name starts with a letter and does not contain spaces or the underscore.

- Remember that R is case-sensitive—so it will think 'Mydata' and 'mydata' are different.

- Don't use the same name for two different things—that will definitely confuse R.

- If R seems to have got itself stuck, or the cursor is a '+' instead of a '>' then you can recover control by pressing the escape 'ESC' key.

- You can use the up arrow key to scroll through previous commands that R has executed; this is handy if you want to submit a command again, because you can modify it before hitting return to resend it to R.

- Every bracket you open, you should close.

- You can solve any problem in R through a little exploration in your search engine of choice.

- Don't give up—you can make R do whatever you want.

Using colours in R

As we discussed in section 1.6, colour can be a powerful tool for data visualization, but it should be used wisely. There is a lot of advice available online on how to select colours for figures, and quite a few people have produced colour palettes for use in R. Here are some websites that might be worth a look:

- Stephen Few's 'Practical Rules for Using Color in Charts':
  http://www.perceptualedge.com/articles/visual_business_intelligence/rules_for_using_color.pdf

- 'What to Consider When Choosing Colors for Data Visualization' uses comparative examples to demonstrate good uses of colour:
  https://www.dataquest.io/blog/what-to-consider-when-choosing-colors-for-data-visualization/

- 'Your Friendly Guide to Colors in Data Visualization' offers links to a range of tools to help in choosing colours:
  https://lisacharlottemuth.com/2016/04/22/Colors-for-DataVis/

- 'How to Choose Colors for Data Visualizations' explains the different types of palette and gives tips and tools for choosing and testing colours:
  https://chartio.com/learn/charts/how-to-choose-colors-data-visualization/

- 'An Alternative to Pink & Blue: Colors for Gender Data':
  https://blog.datawrapper.de/gendercolor/

- 'Top R Color Palettes to Know for Great Data Visualization' presents six palettes and how to use them in your charts:
  https://www.datanovia.com/en/blog/top-r-color-palettes-to-know-for-great-data-visualization/
- 'Colors in R' explains how to specify colours, use predefined palettes, and create your own palette of contiguous colours in R:
  http://www.sthda.com/english/wiki/colors-in-r
- 'Palettes in R' shows how to define palettes and use predefined palettes:
  https://www.r-bloggers.com/palettes-in-r/
- 'Colors and Color Functions' provided by the '`unikn`' package in R:
  https://cran.r-project.org/web/packages/unikn/vignettes/colors.html

Brewer palettes (named after their creator Cynthia Brewer) are colour combinations that have been selected for their perceptual properties and effectiveness at presenting qualitative, sequential, and diverging data. They have become a popular way of choosing appropriate colour palettes for graphics, and now there is even a dedicated '**Rcolorbrewer**' package for use in R. We keep colour selection simple in this book, but if you are interested in more stylish ways to colour figures effectively we encourage you to check out the following:

- 'Brewer Palettes' explains the perceptual properties of Brewer palettes:
  http://mkweb.bcgsc.ca/brewer/
- 'The A-Z Of Rcolorbrewer Palette' presents the different palettes available in this R package:
  https://www.datanovia.com/en/blog/the-a-z-of-rcolorbrewer-palette/
- 'Color Brewer 2.0' allows you to choose a range of colour schemes for use in map data:
  https://colorbrewer2.org/#type=diverging&scheme=RdYlGn&n=8

Discussions about chart junk

We touched on the concept of 'chart junk' in section 1.7, and a quick online search for the term will turn up a range of articles debating what should be considered appropriate usage. If you are keen to learn more about effective figure design, and what you should keep in or leave out, here are a few interesting links that would be a good starting point:

- 'Useful Junk? The Effects of Visual Embellishment on Comprehension and Memorability of Charts', a study by Bateman et al. (2010), found that some visual embellishments—if well designed—aid viewers' long-term recall of the presented data, and questioned the minimalist approach often recommended for figure design:
  http://www.stat.columbia.edu/~gelman/communication/Bateman2010.pdf
- 'The Chartjunk Debate: A Close Examination of Recent Findings' by Few (2011) thoroughly evaluates the methodology and conclusions made by Bateman et al. (2010):
  https://www.perceptualedge.com/articles/visual_business_intelligence/the_chartjunk_debate.pdf
- 'It's Easy to Produce Chartjunk using Microsoft® Excel 2007 but Hard to Make Good Graphs' by Su (2008) highlights some of the issues of using a program

like Excel to produce figures—R is a much better choice for statistical graphing:

https://doi.org/10.1016/j.csda.2008.03.007

- 'Bad Visualizations' is a tumblr page dedicated to graphics people have found where poor or misleading design choices have been made. Some can be frustrating, some can be funny, but they are worth checking out if you're interested in developing your critical thinking about what makes an effective figure:

https://badvisualisations.tumblr.com/

2 PIE CHARTS AND TABLES FOR QUALITATIVE DATA

Learning objectives

By the end of this chapter you should be able to:

- Explain when pie charts and tables are a good choice of data presentation and what design choices you might want to make based on different types of data.
- Produce clear and effective pie charts using nominal data.
- Produce clear and effective pie charts using ordinal data.
- Produce clear and effective tables using Word.

2.1 Introduction: when would you want to use a pie chart or a table?

Pie charts and tables are tools commonly used to visualize qualitative data (also sometimes referred to as 'categorical' data). We defined this sort of data in section 1.3.2 but, briefly, qualitative data describes situations when we have one or more traits that we are interested in, but for these traits individuals are not measured but rather placed in one of several non-overlapping levels of the trait.

Qualitative data can be either nominal (where there is no logical ordering to the levels) or ordinal (where there is an ordering), but in either case the data set will consist entirely of a single value associated with each level of the category. This value might be a count, a percentage, or a single measured value. We will describe examples of each case later in the chapter, but see section 1.3.2 and the online resource 'Data types' from chapter 1 for more on nominal and ordinal data, and further examples.

We believe that pie charts can be misleading, because humans are not as good at comparing the relative areas of pie segments as they are linear representations of data, such as is used in the bar charts discussed in chapter 3. Also, we think data can be expressed more compactly and in an easier-to-assimilate form in a table than a pie chart. However, we appreciate that you might disagree with us; or you might be in a position where you have been specifically instructed to produce such a chart. So, despite our reservations, we will discuss how to produce these charts as effectively as possible. We do this early in this book, because the data sets involved are particularly simple.

The code for pie charts is very easy in R: `pie(x, labels=)`, where `x` will be the data for the area of each slice of the pie and `labels=` will precede a list of the names attributed to each slice. As we outline the steps to producing simple and more refined versions of pie charts in RStudio, this chapter will recap some of the basic data-entering code covered in the 'R Basics' guide from the online resources for chapter 1, but will also cover code that is universal to all figures in R (and therefore will reoccur in later chapters) rather than being exclusive to pie charts.

In the vast majority of cases where you are considering using a pie chart, however, we would strongly recommend that you instead present the data as a bar chart (covered in chapter 3) or a table. Pie charts are perhaps most likely to be useful when your qualitative data is in percentage form and those percentages add up neatly to 100 per cent. Be sure to decide on either a chart or a table, though—there is never any need to present the same data in multiple different formats. Tables are more compact than pie charts and the data contained is easier to assimilate in table form. With pie charts, viewers are also often forced to switch their attention from the data to a legend (aka a key—see section 1.2 for why we use the term 'legend'), with the legend itself also proving problematic if colours are not chosen wisely (see section 1.6 for advice on choosing colours).

Although R can be used to produce tables, we think that Word is instead the best tool for producing clear and visually appealing tables, so we will describe in the second half of this chapter how to go about using Word effectively for this purpose.

 Key point

If you have qualitative data, then the pie charts and tables that we discuss in this chapter should offer good ways of displaying that data.

2.2   Pie charts

2.2.1   Nominal data (unordered categories)

A 2012 study (Advisory Board 2012) investigated which workplace surfaces harbour the most germs, by swabbing surfaces and measuring the adenosine triphosphate (ATP) levels (which indicate contamination by animal, vegetable, bacteria, yeast, and mould cells) from a number of office buildings. The surfaces that were most likely to have ATP levels of 300 or higher (considered a high risk for illness transmission) are outlined in Table 2.1.

Table 2.1 The percentage of the samples from different workplace surfaces that were found to have ATP levels of 300 or higher (Advisory Board 2012).

Surface	%
Computer keyboards	27
Sink handles in break room	75
Water fountain buttons	23
Microwave door handles	48
Refrigerator door handles	26

2.2.1.1 Simple pie chart

As a reminder, when producing any figure in RStudio we recommend using R scripts (see the 'R Basics' guide from the online resources for chapter 1 for more on this) to type and store your code, with annotations. R scripts can be opened in the top left-hand window of RStudio, making it easy to edit and run your code through the Console. We also cover how to input simple data in the 'R Basics' guide from the online resources for chapter 1, and that is how we will start with this first pie chart.

In Table 2.1, we have nominal data, where there is no logical ordering or chronology to the five categories of workplace surface.

Step 1: We first need to input our data into R, by running code that we write in an R script through the RStudio Console. To do so, we create a list of the percentages in the same order as they are in Table 2.1 and give this list a name—here 'germs':

```
germs<-c(27,75,23,48,26)
```

c() just tells R that the values inside the brackets are collected together to form a list.

Step 2: We also need to create a list of the workplace surface names—here 'surfaces'. This list of surfaces should be entered in the same order as their corresponding percentages in the list 'germs':

```
surfaces<-c("Computer keyboards", "Sink handles", "Water
fountain buttons", "Microwave door handles", "Refrigerator door
handles")
```

Note: Another reason we recommend using R scripts to store code rather than a Word document, for example, is that the punctuation you type in a script automatically matches the format R recognizes. For example, R only likes the " " speech marks that R scripts (and some other programs) use—it will not work if you use the curly " " speech marks Word uses automatically. Using R scripts means we have one fewer rule to worry about when coding names for variables or variable levels—though there are still some general tips to bear in mind (we discussed these in Bigger Picture 1.1).

Step 3: Using these lists, we can now substitute these into the 'x' and 'labels=' settings of the pie chart code mentioned in section 2.1 to produce a simple pie chart, which will pop up in the bottom right-hand window of RStudio under the 'Plots' tab.

```
pie(germs, labels=surfaces)
```

You may need to stretch out the RStudio window containing the plot to see the segment labels more clearly (see the 'R Basics' guide from the online resources for chapter 1 for more on using RStudio). This simple pie chart (which we reproduce in Figure 2.1) took very little code to produce. However, as this data is nominal and the categories are not constrained to any particular order, we can improve the chart by ordering the pie segments by size and neatening up where the segments start from. Arranging the categories so that the segments grade sequentially from the largest value (i.e. highest percentage) category to the smallest value (smallest percentage) category is the most intuitive way to display the information, so this is what we do next.

Step 4: Rearranging the segments by size is easily achieved by changing the order in which we input both the lists from earlier. For the list of labels we

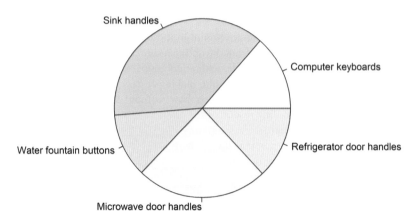

Figure 2.1 A simple pie chart showing the percentage of samples where different workplace surfaces were found to have ATP levels of 300 or higher (Advisory Board 2012), from Table 2.1.

simply create a new **surfaces** list, arranging the labels by descending percentage values:

```
surfaces<-c("Sink handles", "Microwave door handles",
"Computer keyboards", "Refrigerator door handles",
"Water fountain buttons")
```

There are then two ways we could rearrange the percentage values themselves. The first is to simply retype the values into a list called '**germs**' in descending order:

```
germs<-c(75,48,27,26,23)
```

Or alternatively you can create a new list called '**germs**' using the already existing list and use the **sort** function. **sort** tells R we want the values in a particular order, and in this case we want a decreasing order (**decreasing=TRUE**). This may be particularly useful when you have a lot of values, or values with a lot of significant figures:

```
germs<-sort(germs, decreasing=TRUE)
```

Also, if you have a large number of values with a lot of associated labels, a quicker way to rearrange the labels to match their corresponding ordered values can be found in Bigger Picture 2.1. Whichever method you follow, running the code to redefine what the name '**germs**' represents will overwrite the previous list called **germs** that R has stored. You can check this in the top right-hand window of RStudio under the 'Environment' tab, as there will be only one version of your **germs** list stored, and it will match whichever line of code you have run through the Console most recently.

Step 5: Now that we have our values in decreasing order, we will also make our segments run clockwise starting at '12 o'clock'. This is because it is more intuitive for viewers to have the largest slice at the top and then the slices descending in size order as you follow around the chart. By default, R starts the first segment at 0 degrees (i.e. '3 o'clock'), but for a '12 o'clock' start at 90 degrees we can include **clockwise=TRUE** in the simple pie chart code from before:

```
pie(germs, labels=surfaces, clockwise=TRUE)
```

Bigger Picture 2.1
Quickly rearrange values and labels

This is a handy technique that we find useful when organizing our data in R in lots of contexts, not just when graphing. This tip is perhaps particularly useful when you have two rather long lists of values and associated labels. If these come from a data set you have loaded in, you should be able to lift them from the data set without typing them in manually (see later chapters), but here we will give an example starting with our lists of unordered values (**germs**) and labels (**surfaces**) from section 2.2.1.1:

```
germs<-c(27,75,23,48,26)
surfaces<-c("Computer keyboards","Sink
handles", "Water fountain buttons",
"Microwave door handles", "Refrigerator door
handles")
```

Next, we assign the surface label names to each of their associated values in the unordered **germs** list using the function **names**:

```
names(germs) = surfaces
```

If you look at '**germs**' under the 'Environment' tab of the top-right window of RStudio now, you'll see that rather than the list being classed as just numbers ('num') it now states that the list consists of named numbers ('Named num'); that is, each value is labelled with the appropriate surface name. As in the main text, we can now sort the **germs** values, but this will now automatically sort the surface labels to match the reorganization of the values:

```
germs<-sort(germs, decreasing=TRUE)
```

Now we can name our reordered list of labels as '**surfaces**' again, telling R to use the '**names**' from the **germs** list as they are organized now:

```
surfaces<-c(names(germs))
```

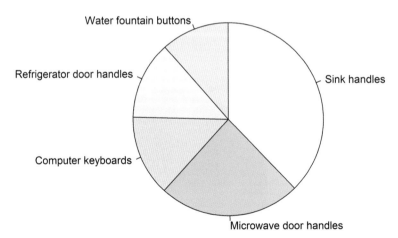

Figure 2.2 As Figure 2.1 but with the categories arranged clockwise from a '12 o'clock' starting point by decreasing value—this is a more intuitive design for readers to follow.

2.2.1.2 Refined pie chart: adding colour, relative percentages, a legend (key), and a chart title

While our simple pie chart (reproduced in Figure 2.2) looks reasonable, it is not particularly attractive to look at. For one, the default colours R has chosen are rather drab and dreary. Let's start by jazzing them up a bit.

Step 1: Using the same ordered lists, **germs** and **surfaces**, from section 2.2.1.1, we can run the code below through the Console to cheer up the pie chart:

```
pie(germs, labels=surfaces, clockwise=TRUE,
col=rainbow(length(surfaces)))
```

col= can be used to specify any colour you want, but if you are listing different colours for various segments you need to make sure that the number of colours you list (e.g. **col=c("red", "blue", "cyan", "magenta", "yellow")**) is equal to the number of segments you need to have colours for. In this case, we are saving ourselves the bother of choosing individual colours and typing them all out by using the **rainbow** function so R automatically fills in a colour palette for a given number of variables: **(length(surfaces))** tells R we want a different colour for each surface type in our data set.

Note: There are more colours than you will ever need available in R!

See section 1.6 for advice on choosing colours effectively and the 'Colour Guide' from the online resources for chapter 1 for a one-page guide to colours we frequently find ourselves using. Or you can find a full list of colours in R online at http://www.stat.columbia.edu/~tzheng/files/Rcolor.pdf.

This has already gone some way to improving the visual appeal of our pie chart, but the labels for each segment look a bit messy. A neater way to present them is to use a legend (aka a key, but see section 1.2 for a reminder of why we use the term 'legend').

Step 2: As we're moving the long segment labels to a legend, we might instead want to add new labels to the chart that clearly show the percentage value each segment represents. We can do this with the code:

```
pie(germs, labels=paste(germs,"%"), clockwise=TRUE, col=
c("darkgoldenrod1", "orchid3", "springgreen3","yellow2",
"royalblue3"))
```

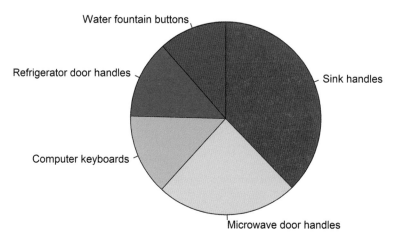

Figure 2.3 As Figure 2.2 but with much more eye-catching colours.

The main difference to our code here is in the **labels** argument. Rather than just naming labels with the list of values, our pie chart will look smarter if each value is followed by the % symbol. To present values alongside text symbols in R we can use the **paste** function. Its use along with our **germs** list of values tells R that we want our labels to consist of each value of the **germs** list in turn, each followed by the % symbol. See section 6.3.3 for another use of **paste** to present values alongside text, and section 8.6 and Scientific Approach 8.1 for more details on the **paste** function (and its use alongside the **expression** function).

Notice also that this time we have chosen our own list of (arguably more muted and visually appealing) colours with the **col** argument. This gives us the incomplete Figure 2.4–incomplete as there is currently no explanation as to what the different slices represent now.

Here, the values we were dealing with were listed as percentages in Table 2.1 but the percentages were not relative to each other: rather they described the percentage of the samples where each work surface was found to have an ATP level of 300 or higher. The percentages in Figure 2.4 consequently do not neatly add up to 100 per cent. It might, therefore, be interesting to consider the likelihood of these top five work surfaces being contaminated relative to each other. Use of a pie chart also makes more sense if the pie's various parts add up to a whole 100 per cent—in fact, some people argue that pie charts should never show raw numbers at all, only ever the relative percentages. We can do this by calculating the percentage of the pie chart each segment should take up relative to the other segments:

```
piepercent<-round(100*germs/sum(germs),1)
```

In this code, we are getting R to do all our calculating for us. We create the new list of values (**piepercent**) and tell R that for each value in our **germs** list we want it to calculate 100 multiplied by that value and then to divide this result by the sum total of all the different values in the **germs** list. We also here use the function **round** to tell R to round the values represented ahead of the comma in its containing brackets to the nearest 1 decimal place (stated after the comma in the containing brackets). Sometimes you will want to keep the values your pie chart shows as those of the original data set, but this step is useful where you want the relative proportions of each recorded response to be clear. We have done so here to demonstrate the code, but for a data set of this sort we

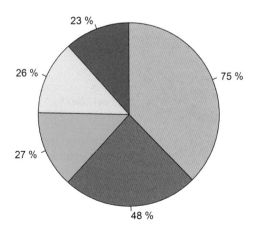

Figure 2.4 As Figure 2.3 but with the pie chart labels now listing the percentage values from Table 2.1 rather than the workplace surface names. The colours used were also specified in the code. Without an explanation of what the different coloured segments represent, this is an incomplete pie chart.

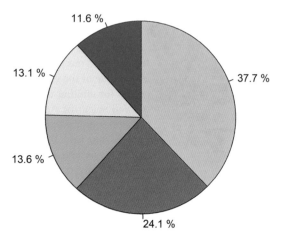

Figure 2.5 As Figure 2.4 but with the pie chart labels now listing the percentage values of the likelihood of each workplace surface being contaminated relative to the other surfaces represented. That is, if we find that a sample from a workplace surface is contaminated, these percentages describe the relative likelihoods that this sample comes from one of five (currently unspecified) work surface types. Without an explanation of what particular surfaces the different coloured segments represent, though, this is still an incomplete pie chart.

would strongly advise producing a table (covered in section 2.3) or bar chart (covered in chapter 3) instead.

Producing our pie chart again, we can substitute the **germs** list with our new **piepercent** list to get Figure 2.5:

```
pie(piepercent, labels=paste(piepercent,"%"), clockwise=TRUE,
col= c("darkgoldenrod1", "orchid3", "springgreen3","yellow2",
"royalblue3"))
```

Again, if you are interested in the use of **paste** to present values alongside text or any particular mathematical symbols, we recommend you look into section 8.6 and Scientific Approach 8.1 for details.

Step 3: Whether we have presented the values from the original data set or converted these into relative percentages, we can now add the segment names (here, our work surfaces) in a legend, making sure we using the same list of colours in the same order:

```
legend("bottomright",legend=surfaces, cex=0.7,fill=
c("darkgoldenrod1", "orchid3", "springgreen3","yellow2",
"royalblue3"))
```

'**bottomright**' tells R where we want the legend positioned relative to the chart (see section 7.4 for other ways you can position and customize your legends), **legend** is where we tell R the list of names we want included in the legend (here, our existing list '**surfaces**'), **cex** is the font size (you can play about with this, but 0.7 is good in this case for fitting the legend next to our chart), and **fill** is a list of the colours for the legend that will correspond with the chart.

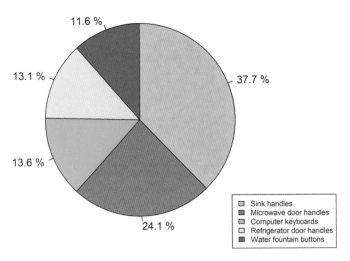

11.6 %

13.1 %

37.7 %

13.6 %

24.1 %

☐ Sink handles
■ Microwave door handles
☐ Computer keyboards
☐ Refrigerator door handles
■ Water fountain buttons

Figure 2.6 A high-quality pie chart displaying relative percentages of the data in Table 2.1 (Advisory Board 2012), with specified colours and a corresponding legend.

Step 4: Figures should always be accompanied by a figure caption or a figure title to explain what they show. Any figures included in a document or scientific journal should only be accompanied by a figure caption (we provide advice on writing these in section 1.2 and Scientific Approach 1.2).

However, in cases where viewers may have a harder time reading the small print of a figure caption, you may want to use a title with a larger font. This is most likely to apply when you are preparing a figure for a poster or a projected image. To do this, we simply include the argument `main` in our R script code as below, and run the following through the R Console:

```
pie(germs, labels=paste(germs,"%"), main="Which workplace
surfaces harbour the most germs?", clockwise=TRUE, col=
c("darkgoldenrod1", "orchid3", "springgreen3", "yellow2",
"royalblue3"))
legend("bottomright",legend=surfaces,cex=0.7,fill=
c("darkgoldenrod1", "orchid3", "springgreen3","yellow2",
"royalblue3"))
```

Using the same nominal data set, we have gone from our simple pie chart of Figure 2.1 to our refined, professional-looking Figure 2.7.

2.2.2 Ordinal data (ordered categories)

'Rewilding' is a concept in conservation biology involving the reintroduction of species that are now locally extinct to part of their former range. While rewilding is suggested to restore natural ecosystem processes that have been lost and encourage tourism, some people have concerns about the danger certain species may pose to people and livestock after reintroduction. Attitudes towards rewilding large mammals can range from strong support to strong opposition. The responses of 2083 adults living in the UK in a 2019 survey (YouGov 2019) to the question 'To what extent would you support or oppose reintroducing species of animal to the UK?' are outlined in Table 2.2.

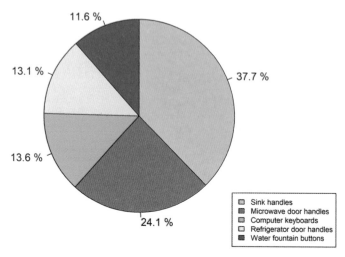

Which workplace surfaces harbour the most germs?

Figure 2.7 As Figure 2.6 but with a chart title.

Table 2.2 In a 2019 survey of 2083 UK adults (YouGov 2019), the percentages of individuals selecting each of five possible responses after being asked 'To what extent would you support or oppose reintroducing species of animal to the UK?'

Response	%
Strongly support	36
Tend to support	46
Tend to oppose	6
Strongly oppose	1
Don't know	11

2.2.2.1 Refined pie chart

We worked through the various steps to creating a simple pie chart for the nominal data example above, so this time we will jump right in and produce a high-quality pie chart for this second example. The data in Table 2.2, though, is ordinal as there is a logical ordering to the response categories—responses range from strong support to strong opposition, and it makes sense for the 'Don't know' category to take last place as it is most easily appreciated having previously seen the alternative available options.

Step 1: We first need to input our data into R, by running code that we write in an R script through the RStudio Console. Because we are working with ordinal data, it is important that we input our data in the same order as the responses. We therefore do not need to worry about sorting the data depending on value. Also, note that the response percentages are already relative to each other, adding up to 100 per cent—this means we also do not need to get R to calculate segment percentages as we did in the nominal data example. It also means that this data is more suitable for presentation in a pie chart than that of the nominal data example. Here we simply create a list of the percentages that we call `percent`:

```
percent<-c(36,46,6,1,11)
```

Step 2: We can also create a list of the responses in the same order as we inputted the percentage values—this will be useful when we add our legend:

```
responses<-c("Strongly support", "Tend to support", "Tend to
oppose", "Strongly oppose", "Don't know")
```

Step 3: Now, using the code we covered with the earlier example, we can easily produce a high-quality pie chart with segments running clockwise, labels of the percentages each segment represents, and colours we have specified (see section 1.6 for advice on choosing colours for ordinal data):

```
pie(percent, labels=paste(percent,"%"), clockwise=TRUE, col=
c("forestgreen", "green","red", "red4", "cornsilk2"))
```

Step 4: Then add a legend using our list of `responses`, specifying those same colours in the same order (see section 7.4 for further legend positioning options):

```
legend("bottomright", legend=responses, cex=0.7,fill=
c("forestgreen", "green","red", "red4", "cornsilk2"))
```

Step 5: If we want to include a chart title, we can do so by including `main` as an argument in the pie chart code. Although remember that most scientific figures are usually accompanied by a detailed figure caption instead of a title (see our advice on writing these in section 1.2 and Scientific Approaches 1.2)—we include `main` in the code below to demonstrate how to do this for cases where you decide a title is more appropriate for your purpose:

```
pie(percent, labels=paste(percent,"%"), clockwise=TRUE, col=
c("forestgreen", "green","red", "red4", "cornsilk2"), main="UK
adult attitudes towards rewilding")
legend("bottomright",responses, cex=0.7,fill=c("forestgreen",
"green", "red", "red4", "cornsilk2"))
```

Our resulting pie chart in Figure 2.8 is clear to read and the ordinal data is still presented logically in order of the various responses. As with any figure in R, there are additional customization options available for pie charts, including variable shading of segments and 3D, and we provide details on these in the online resource for this chapter 'Shading and 3D' and its accompanying 'R script for Shading and 3D'. However, we feel that these design features will rarely, if ever, increase the ability of your graphic to communicate the data effectively (and, indeed, can often reduce the clarity of figures).

2.2.3 Tips and warnings for pie charts

Here are some additional top tips for producing pie charts:

> *Tip 1*: Try not to use a pie chart if there will be more than seven segments. If you have more than this, then consider whether any segments can be combined, as long as these composite categories are carefully explained to the reader (see Tip 4).

> *Tip 2*: Make sure that segments start from the uppermost point.

UK adult attitudes towards rewilding

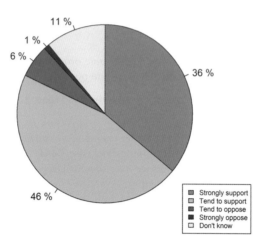

11 %
1 %
6 %
36 %
46 %

Strongly support
Tend to support
Tend to oppose
Strongly oppose
Don't know

Figure 2.8 A high-quality pie chart of ordinal data, showing the percentage responses of 2083 UK adults when asked about their attitudes towards rewilding in a 2019 study (YouGov 2019), from Table 2.2.

Tip 3: Avoid using cut-outs, 3D, or any other graphic design feature that will distort the value and perceived importance of any of the segments.

Tip 4: Be very slow to lump 'inconvenient' data into a mystery 'All others' slice. Abusing pie chart categories in this way may hide valuable data.

Tip 5: Do not exclude data that complicates your story—this will make the remaining slices of the chart grow and distort how your data is interpreted. Make sure to show all of your data, even if it does not fit as nicely as you had hoped to the story you would like to tell.

Tip 6: Consider whether a table might be the more effective way to present your data (we cover how to produce effective tables in section 2.3). This may be especially important if there are only small differences in the values of your data items.

💡 Key point

We generally prefer a table to a pie chart. But if you disagree with us then the advice in this section should help you easily to produce as effective a pie chart as possible.

2.3   Tables

Tables are a great choice when you want a reader to be able to consult specific values easily; but if you want to clearly visualize a **trend** or relationship between levels of a qualitative variable or between variables, then you may want to consider a graph instead (a bar chart is a good choice if your data is qualitative—we cover these in chapter 3). Generally, we would advise that a table with four or fewer values is a waste of space; if you can give all the numbers in a single easy-to-read sentence, then do that rather than construct a table. If you have enough values to present, though, as mentioned in section 2.1, we believe that tables are (almost always) a superior presentation tool for qualitative data, compared with pie charts.

This is due to tables being compact and we feel that the data contained is easier to assimilate. Also, due to the presentation of values as unequal segments, the more data values you add to a pie chart the messier and harder to assimilate it gets. In contrast, any cell of a table is as readable as the next (no matter the size of the table).

Although you can produce tables using R, we think that Word is the best software for producing high-quality tables (we detail some of its main table features in section 2.3.2) and this is the software we use for the nominal and ordinal data examples in sections 2.3.3–2.3.5). First, though, we set out some general guidelines for producing highly effective tables, that should apply whether Word is your tool of choice or not.

2.3.1 Guidelines for producing tables

We think that decisions surrounding the presentation of data in tabular form fall under three main categories: how the data is organized and presented, how the table headings are formatted, and how the table overall is presented. For this reason, below we have separated out the guidelines we follow when producing tables (and our top tips) under these subheadings. If you struggle to make sense of any of these guidelines, then we talk through the thinking behind some tables in sections 2.3.3–2.3.5 and in the online resource 'Authors' attempts for chapter 2'; and of course you would hope that all the tables in this book should conform to our guidelines.

2.3.1.1 Data organization and presentation

- Always ensure that all your numerical data are in the same units (where possible), rounded appropriately, and have the same number of decimal places (if applicable).

- Arrange the categories of the data that you want the user to compare in columns—vertical comparison is easier than horizontal comparison, so make sure that numerical values run downwards rather than across the page wherever possible.

- With nominal data, it is easiest for readers to draw comparisons between values when they are increasing or decreasing as they move down the table.

- With ordinal data, the categories of the nominal variable must be kept in their logical order, rather than ordering them by corresponding value sizes.

- Whether you are dealing with nominal or ordinal data, put some thought into how you group and order items for clear interpretation.

- If the table includes time as a variable, present time running left-to-right or top-to-bottom.

- Right-align data in columns to ensure that digits of equal value lie under similar valued digits in all numbers that are to be compared. Numerical data is read right-to-left and it is easier to compare numbers if the units (tens, hundreds, thousands) are aligned in the same columns.

- Left-align text, because textual data (if it is in English) is read from left to right.
- Only embolden values within tables if such emboldening is informative and important for highlighting key data (e.g. perhaps highlighting significant p-values in a table of statistical testing results). Especially do not embolden if emboldening numbers moves them out of the correct alignment of units within columns (you could, alternatively, change colour or underline values you want to draw particular attention to).
- Add totals, where they are useful to interpretation, to the right-hand side of rows and/or the bottom of columns of data.

2.3.1.2 Heading formatting

- Provide informative column (and row, if applicable) headings. You need to give units, sample sizes, and errors where necessary, using an abbreviation and a footnote if you have to.
- Any footnotes that need to be added to further explain a particular variable should be put against the column or row heading rather than a data cell (because this would mess up the unit alignment and make comparison of data more difficult).
- Clearly differentiate headings from the data, but do not make them so eye-catching that they distract from the data. Emboldening alone is usually sufficient—there is no need to change the colour or underline, for example, as well.
- Do not capitalize every word in column or row headings, as unnecessary capitalization slows down reading and interpretation.
- Align column headings to match the alignment of their data in the table—this makes it clear which data the headings are associated with and keeps the vertical lines of the table clean.

2.3.1.3 General table presentation

- Reduce the use of cell borders where possible—they can add clutter and disrupt a reader's scanning down or across when trying to compare values. Use horizontal lines only, and sparingly at that. Lines are useful, though, at the top and bottom of tables to give them structure, and perhaps to the left and right of row headings if applicable.
- Reduce clutter and redundancy: if all elements in a table are in the same units, for example, then explain this in the table caption, not repeated for every column.
- For larger tables, consider deleting rows or columns that you never discuss in the text.
- For really large tables, you might want to consider splitting one table into multiple tables that are easier to interpret—but try to avoid duplicating data across tables.
- If you have multiple tables, be consistent between them in formatting, ordering, and nomenclature.

- Consider using alternate shading of rows to assist with the reading of multi-column tables. When you have a lot of columns, alternate shading can really help readers follow a specific row across, but this isn't necessary and perhaps adds clutter in simple tables. However, if you want to see what this looks like, then see Graeme's solutions in the online resource 'Authors' attempts for chapter 2', and Rosalind's large multi-column table in 'Authors' attempts for chapter 5'. Key to shading, though, (whether done as alternate shading or general background colours) is that you carefully select background colours that contrast highly enough with your text and data so that they can be easily read (we mentioned this, among other general concepts in colour choice, in Scientific Approach 1.3); adding in colour should not come at the cost of your table's accessibility.

- Give tables an informative table caption that makes the table intelligible without reference to the text, but that does not contain extensive interpretation of the data. Table captions are often placed above, rather than below, the table, but you will find both variations in scientific publications. You may also find table captions referred to as table 'legends', but we use 'caption' here for consistency with our language around figures (see section 1.2 for a recap of why we use 'figure caption' instead of 'figure legend').

- Refer to all tables in the text and be consistent in table number, row, and column names between tables and text.

2.3.2 Producing tables in Word

As mentioned in section 2.3, we think that Word is the best program with which to produce highly effective tables.

To draw a basic table in Word, you simply open the 'Insert' ribbon on the toolbar at the top, click on the 'Table' button, and drag the mouse down over the required number of cells for rows and columns. A good rule to stick by is to create a separate cell for each item—this helps you avoid layout problems arising from resizing or missing values. You can also then pull in the sides of the columns or rows to resize them and minimize blank space.

Once you have filled your table in with your data (following some of the basic data organization and heading formatting guidelines we set out in sections 2.3.1.1 and 2.3.1.2), you can then use the many tools available in Word to improve the presentation and ease of interpretation. Below, we direct you to some of the most helpful features that you might use in order to follow the guidelines from section 2.3.1, where appropriate for your data.

Many of the features are accessible via right-click shortcuts, which you might choose to use once you are more familiar with producing tables in Word. Here, though, we will describe how to access tools using the main toolbar at the top of Word documents. To apply the following features, first highlight the cells of the table that you want the feature to apply to then follow the steps below:

- To align table columns, either hold Control + R (to right-align) or Control + L (to left-align), or click on the 'Align Right' or 'Align Left' button on the 'Home' ribbon of the toolbar (to find these, hover over the four adjacent alignment buttons—consisting of different arrangements of four short horizontal lines).

- To embolden column/row headings or key values (though see our warning about emboldening in section 2.3.1.1), click the 'Bold' button (a bold, capital 'B') on the 'Home' ribbon of the toolbar.

- To edit the default borders around cells in the table (a good idea to reduce clutter, see section 2.3.1.3), open the 'Table Design' ribbon of the toolbar, click the drop-down arrow on the 'Borders' button, select 'Borders and Shading', and then use the preview diagram to add or remove borders from around the cell as needed.

- To add shading to particular rows or columns, open the 'Table Design' ribbon of the toolbar, click the drop-down arrow on the 'Shading' button, and select the desired background colour (though bear in mind our advice from section 2.3.1.3 regarding use of colour).

- To add a footnote number or letter against a heading, use the 'Superscript' button (looks like x^2) on the 'Home' ribbon of the toolbar.

- If at any point you decide you need to add an extra row or column to your table (perhaps to add a footnote, or a useful 'Total'), highlight the row or column you want it to appear beside, open the table-specific 'Layout' ribbon at the right-hand side of the toolbar, and then click one of the 'Insert Above', 'Insert Below', 'Insert Left', or 'Insert Right' buttons as needed.

- To combine two or more existing cells of a table (perhaps useful if adding a footnote), open the table-specific 'Layout' ribbon at the right-hand side of the toolbar and then click on the 'Merge Cells' button.

- If at any point you decide you need to delete a particular row or column, open the table-specific 'Layout' ribbon at the right-hand side of the toolbar, click the 'Delete' button, and then select the appropriate option from the drop-down menu that appears.

Word does also provide a selection of 'Table Styles' you might want to consider using. Once you have selected your table, you can click on the 'Table Design' ribbon of the toolbar, and hover over the different 'Table Styles' to see their formatting applied to your table (click on the arrows to the right of the 'Table Styles' box to scroll through the whole list). To select a style, simply left-click on it. You can also filter the 'Table Styles' by ticking or removing ticks from the options given on the left-hand side of the 'Table Design' ribbon, e.g. whether you want your table to have a distinct 'First Column' or 'Banded Rows' (aka alternate shading: see our advice on this in section 2.3.1.3). Sometimes these automatic 'Table Styles' will be useful to you, but if you choose to use one you should always check that it is presenting the data as effectively as possible, and isn't overcomplicated, cluttered, or too 'style-over-substance'—you can always edit specific features after applying a style if needed.

In the following examples (sections 2.3.3–2.3.5) we have created tables from the ground up, without selecting a default 'Table Style', working through the guidelines set out in section 2.3.1 to ensure that our data is presented clearly and effectively.

2.3.3 Nominal data (unordered categories)

Our first data set contains more than four values, making a table suitable, but will fit into a simple two-column table. The data concerns the top carbon dioxide (CO_2) emitting countries, based on 2018 data from the European Commission's Emissions Database for Global Atmospheric Research (BBC News 2020). The megatonnes of CO_2 released per year by those countries are: Canada,

594; China, 11256; EU, 3457; Germany, 753; India, 2622; Indonesia, 558; Iran, 728; Japan, 1199; Russia, 1748; Saudi Arabia, 625; South Korea, 695; US, 5275. The data is nominal, as there is no single most logical order of the countries.

We produced Table 2.3 using Word (see section 2.3.2 for some of the key tools we used) and by following the guidelines set out in section 2.3.1. Several key design choices specific to this data were made when producing Table 2.3. Firstly, as this is nominal data, it is easiest for readers to draw comparisons between values when they are increasing or decreasing as they move down the table. Therefore, we ordered the countries by decreasing emissions of CO_2 per year. Secondly, we decided that it would be useful for us to specify what a megatonne consists of; but as this explanation would take up a bit of space, we decided to include this as a footnote. We did this by inserting a new row below the table for the explanation, adding superscripted numbers after the word 'Megatonne' in the column heading and at the start of the footnote explanation, and merging the cells in the bottom row so that our footnote could run the width of the entire table. Probably the most striking aspect of this table (and all our tables), compared to many that you will see, is the lack of lines. We think lots of lines distract your eye from the data; so we use no vertical lines on cell boundaries, and only use horizontal ones sparingly.

2.3.4 Ordinal data (ordered categories)

A 2018 report by Ipsos MORI looked into public attitudes to animal research. In response to the statement 'It does not bother me if animals are used in scientific research', across 1011 British adults 15 per cent opted for 'Agree', 18 per cent for 'Neither agree nor disagree', 66 per cent 'Disagree', and 1 per cent 'Don't know'. This is ordinal data, as there is a clear logic to the ordering of the responses, moving from agreement to disagreement and finally uncertainty.

Table 2.3 The world's top emitters of CO_2 in 2018. Data originally sourced from EC Emissions Database for Global Atmospheric Research (BBC News 2020).

Country	Megatonnes[1] of CO_2 per year
China	11256
US	5275
EU	3457
India	2622
Russia	1748
Japan	1199
Germany	753
Iran	728
South Korea	695
Saudi Arabia	625
Canada	594
Indonesia	558

[1]One megatonne = 1,000,000 tonnes

Table 2.4 The responses of 1011 British adults in 2018 when asked to respond to the statement: 'It does not bother me if animals are used in scientific research' (Ipsos MORI 2018).

Response	%
Agree	15
Neither agree nor disagree	18
Disagree	66
Don't know	1

You would not, for example, organize the responses as: 'Agree', 'Don't know', 'Disagree', 'Neither agree nor disagree'. Although there are only four categories of responses here, and we would advise that producing a table of four data values may be a waste of space in a scientific report or paper, we will use this data as an example because there may be some cases (such as when producing posters or giving a presentation) where a table of such a simple data set would help to quickly communicate and visualize your findings.

As with Table 2.3, we produced Table 2.4 using Word (see section 2.3.2 for key table-related features in Word) and by following the guidelines set out in section 2.3.1. As this is ordinal data, though, a key difference from Table 2.3 is that the categories of response had to remain organized in their logical order, rather than the rows being ordered by increasing or decreasing values.

2.3.5 Large data sets

For larger data sets, such as those involving subcategories of multiple different categories (that could not be presented in a single pie chart), clear presentation in table form is even more essential. The bigger a table is, the harder it is for readers to understand, so we have to be thoughtful about how we present it. The ordinal data set we will use to explore large tables concerns a survey of Americans' views on human creation in relation to education, religious preference, and church attendance from a 2017 study (Gallup 2017). Participants were asked which of three statements came closest to their views on the origin and development of human beings. Table 2.5 outlines the percentage of responses given by participants occupying the various subcategories of the three main categories of 'Education', 'Religious Preference', and 'Church Attendance'.

Again, Table 2.5 was produced using Word (the key tools used are included in section 2.3.2) and by following the guidelines set out in section 2.3.1. The categories and subcategories involved in this data meant that organizing rows thoughtfully was essential to facilitate easy interpretation and comparison. The subcategories could all be logically-ordered to some extent, making the data ordinal (while such ordering might be less obvious for the 'Religious Preference' category, it would have seemed odd to position the 'None' option in between the two Christian options). Because of this, the rows within each category were ordered following the logical order of the subcategories rather than by the values for any of the responses. To distinguish the three categories (Education, Religious Preference, and Church Attendance) from their multiple subcategories, beyond positioning them in separate rows, we used shading, carefully choosing a colour that still had good contrast with the text and would

Table 2.5 The results of a 2017 study asking Americans of various different educational and religious backgrounds for their views on human creation (Gallup 2017). Values give the relative percentage of the views given by participants from each background.

	Humans evolved, God guided process	Humans evolved, God had no part in process	God created humans in present form within last 10,000 years
Education			
High school or less	33	12	48
Some college	38	16	42
College graduate	45	27	24
Postgraduate	45	31	21
Religious Preference			
Protestant/Other Christian	39	6	50
Catholic	45	11	37
None	29	57	9
Church Attendance			
Weekly	28	1	65
Nearly weekly/Monthly	44	6	45
Less often	40	35	21

print well in black-and-white. Alternatively, we could have emboldened the category titles 'Education', 'Religious Preference', and 'Church Attendance'. However, to us, shading here usefully structured the table, allowing for easy reference to specific categories without impeding comparison of values or distracting too much from the data.

Another feature to note in Table 2.5 is that, because all the numeric data is in the same units (relative percentages), this unit is explained in the table caption, rather than being repeated for every column. What constitutes the perfect presentation for a particular table is ultimately subjective, and may vary depending on the data you are presenting, but as long as you keep readability and ease of interpretation at the front of your mind (using the guidelines from section 2.3.1 as a guide) you will be able to produce effective and professional tables.

 Key point

Word is a good tool for producing highly effective tables, but no matter what you use to produce your table we offer advice in this section that should help you make your tables more effective.

2.4   Stretch your understanding

If you want to cement and stretch your understanding of the concepts and methods we have outlined in this chapter, have a go at making clear and effective pie charts and tables from these extra data sets.

Where does UK beach waste come from?

This data set is based on a Marine Conservation Society survey of 339 beaches across the UK (BBC News 2017):

- 46.2 per cent unknown
- 30.4 per cent public
- 10.8 per cent fishing
- 8.5 per cent sewage-related debris
- 2.9 per cent shipping
- 1 per cent fly-tipped
- 0.2 per cent medical.

Attitudes to alcohol

A 2017 study into public attitudes to alcohol asked 1700 English and Scottish drinkers whether 'People in Scotland/England are generally discouraged or encouraged to drink alcohol' (Li et al. 2017):

- 62 respondents thought that people are 'Strongly discouraged'
- 191 respondents thought that people are 'Discouraged'
- 603 respondents thought 'Neither'
- 512 respondents thought people are 'Encouraged'
- 332 respondents thought people are 'Strongly Encouraged'

For both data sets, think about whether the data is nominal or ordinal, and how it can most effectively be presented. For the second example, you may want to consider transforming the original data values into percentages.

Once you've had a go yourself, check out the online resources to find our individual attempts at these exercises, including the code we used ('R script for Authors' attempts chapter 2') and the finished charts ('Authors' attempts for chapter 2').

Chapter Summary

- Pie charts and tables can be used to visualize qualitative (aka categorical) data, be it nominal (no logical ordering of categories) or ordinal (with a logical order to categories).
- Pie charts can be tricky to interpret accurately, and often a table could express the same information in a more compact and easier-to-assimilate form. Bar charts are also often a better graphical choice than pie charts, as humans are better at reading values off a linear scale (we discuss qualitative bar charts in chapter 3).
- Pie charts are most likely to be useful when you have qualitative data in percentage form and that data adds up to 100 per cent.

- Using R, the basic code for pie charts is: `pie(x, labels=)` but additional code can be used to improve the clarity of the data.
- Segments of pie charts are most easily interpreted when: i) they ascend or descend in size order as you move around from the top (nominal data), or ii) they follow the logical ordering of the categories (ordinal data).
- Colour in pie charts is always informative, but a legend is essential and colours should be chosen carefully.
- Word is a great program for producing high-quality tables—its many features, including emboldening and removal of cell borders, make it easy to produce clear and visually appealing tables.

Online Resources

The following online resources are available for this chapter at www.oup.com/he/humphreys-obp1e:

- R script for chapter 2
- Authors' attempts for chapter 2
- R script for Authors' attempts chapter 2
- Shading and 3D
- R script for Shading and 3D

Further Reading

If you are interested in learning more about why we advise against the use of pie charts, we recommend the following links:

- 'Graph Comprehension: An Experiment in Displaying Data as Bar Charts, Pie Charts and Tables with and without the Gratuitous 3rd Dimension':
  https://www.rand.org/pubs/working_papers/WR618.html
- 'Save the Pies for Dessert':
  http://www.perceptualedge.com/articles/visual_business_intelligence/save_the_pies_for_dessert.pdf

Some advice on when tables are often better used than graphs is given here:

- 'Do You Know When to Use Tables vs. Charts?':
  https://infogram.com/blog/do-you-know-when-to-use-tables-vs-charts/

3 BAR CHARTS FOR QUALITATIVE DATA

Learning objectives

By the end of this chapter you should be able to:

- Explain when a bar chart is a good choice of data presentation and what design choices you might want to make based on different types of data.
- Produce clear and effective bar charts for both **nominal** and **ordinal** data.
- Produce clear and effective grouped bar charts, allowing easy comparison between groups.
- Produce clear and effective horizontal bar charts, where appropriate for the data.
- Customize bar charts by changing such aspects as bar order, colour, and axis features.
- Understand cases where it is (and is not) appropriate to make y-axes start at non-zero values.

3.1 Introduction: when would you want to use a bar chart?

Bar charts are an effective way to visualize qualitative data, also sometimes referred to as 'categorical' data (see section 1.3 for discussion of different data types). All of the qualitative data examples that we encountered in chapter 2 on pie charts could also be presented as bar charts. Indeed, as you work through this chapter we hope you see that most of the weaknesses of pie charts that we discussed in chapter 2 can be avoided by using a bar chart instead. Bar charts can also allow easier comparison across several similar qualitative data sets than a series of separate pie charts. However, you should still ponder whether presenting some data as a bar chart or as a table might be best for your report. One way to look at this choice is that if you want the reader to appreciate the absolute values then this should cause you to lean towards a table; but if a trend or distribution in relative values across categories is what you want to show then the bar chart might be ideal.

The basic code for bar charts is easy in R—essentially: `barplot(x, names.arg=c())`—but we will demonstrate ways to make bar charts more visually appealing with some extra code. Some of the R coding we will use builds on what we learned in chapters 1 and 2, but we will introduce you to some new twists too.

Key point

Bar charts often offer a more effective way to present qualitative data than a pie chart, and this becomes ever more stark the more complicated your data set.

3.2   Nominal data (unordered categories)

A survey of small mammal abundances was conducted along a series of newly introduced hedgerows. Table 3.1 shows the species of the first 200 small mammals trapped as part of the survey.

3.2.1 Simple bar chart

The data in Table 3.1 is an example of nominal data, because there is no single most logical ordering to the five small mammal species (see section 1.3 for a full discussion of data types). We described how to enter simple data into R in the 'R Basics' guide in the online resources for chapter 1, and this data set is small enough to do this easily.

Step 1: We can very quickly input our data into R by creating a list of our values and giving that list a name, here 'abundance':

```
abundance<-c(53,15,76,39,17)
```

Step 2: Using this list, we can now create a simple bar chart easily with the code:

```
barplot(abundance, names.arg=c("Common shrew", "Pygmy shrew",
"Field mouse","Field vole", "Bank vole"),ylab= "Abundance",
xlab="Species")
```

The `names.arg` function simply tells R that we want our bars to be labelled with the following list of names. `ylab` tells R the y-axis label and `xlab` tells R the x-axis label.

This simple plot (reproduced in Figure 3.1) is not a bad start. However, at a glance it might be a bit difficult to determine whether more pygmy shrews or bank voles were trapped. Where you have nominal data such as this in a bar chart, arranging the categories so that the bars grade sequentially from the largest value (i.e. most abundant) category to the smallest value (least abundant) category is the most intuitive way to display the information, helping the reader to easily interpret the data.

Table 3.1 The recorded abundances of the first 200 small mammals trapped during a survey along newly introduced hedgerows.

Species	Abundance
Common shrew	53
Pygmy shrew	15
Field mouse	76
Field vole	39
Bank vole	17

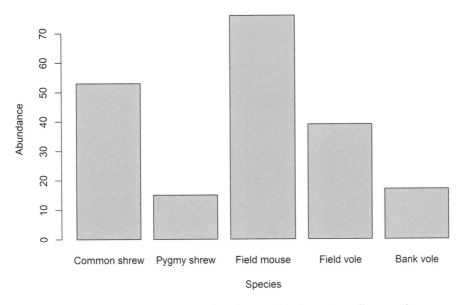

Figure 3.1 A simple bar chart showing the abundances of the first 200 small mammals trapped in a survey conducted along a series of newly introduced hedgerows (data from Table 3.1). The bar for 'Field mouse' extends beyond the y-axis and the exact values of the bars are difficult to estimate.

Step 3: We can easily rearrange the bars by height, by changing the order in which we input our data and x-axis labels. One way (we discussed others in chapter 2) is to create a new list called '**abundance**' from the existing list and use the **sort** function, telling R we want the values in decreasing order. This may be a particularly useful shortcut (as opposed to retyping all the data) when you have a lot of values, or values with a lot of significant figures (but remember to reorder the names to match):

```
abundance<-sort(abundance, decreasing=TRUE)
```

This will overwrite the previous list called '**abundance**' that R had stored. Now we can have a look at this plot:

```
barplot(abundance, names.arg=c("Field mouse","Common shrew",
"Field vole", "Bank vole","Pygmy shrew"), ylab= "Abundance",
xlab="Species")
```

This looks better, but our graph would look even neater if we set the y-axis maximum to 80 (a value just higher than our largest frequency).

Step 4: We can easily include set limits for our y-axis with the code:

```
barplot(abundance, names.arg=c("Field mouse","Common shrew",
"Field vole", "Bank vole","Pygmy shrew"), ylab= "Abundance",
xlab="Species",ylim=c(0,80))
```

ylim here tells R the bounds we want the y-axis to run between, in this case from 0 to 80. We now have a very effective and clear, simple bar chart (reproduced as Figure 3.2).

By offering us a linear arrangement of rectangles, you should be able to see that a bar chart avoids many of the optical illusion issues with pie charts. Also,

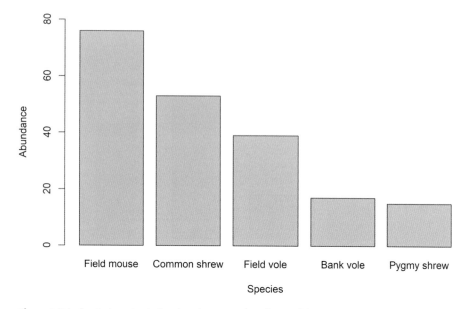

Figure 3.2 A simple bar chart showing the same data from Table 3.1, but with the bars in descending order of abundance and a y-axis that covers the full range of our data.

each bar is allocated the same space of the x-axis, so we do not have the same problems with seeing low-valued categories as we can have with a pie chart. Lastly, we have a lot more space to easily label each bar than we do with pie charts. You are much better off with a bar chart than a pie chart; but stop to consider whether the bar chart in Figure 3.2 is better than a rearranged version of the table that we started with.

3.2.2 Refined bar chart: adding grid lines and colour

Not all figures need grid lines (sometimes called 'graph lines') behind the data they present, but they can be used to good effect where the precise values of the data presented may not be easily discernible otherwise. We may want to add horizontal grid lines behind the bars of our bar chart to make the values each bar represents more readable, aiding comparison—in particular, it may be useful for us to make clearer the difference between the values for 'Bank vole' and 'Pygmy shrew', which appear fairly close together (have a look at Figure 3.2). We can achieve this in two stages: first we produce a simple plot and add our grid lines, and then we then sit a final, clear bar chart on top of those grid lines.

Step 1: Let's plot a simple bar chart with the y-axis defined as above—there is no need to include all the details of our final plot yet:

```
barplot(abundance, ylim=c(0,80))
```

Step 2: Looking at the stripped-down plot this produces in R, we can see that it might aid interpretation if we add some lighter horizontal grid lines at every '2' up the y-axis and some darker lines at every '20'. We can draw these using `abline`:

```
abline(h=(seq(0,80,2)), col="lightgray")
abline(h=(seq(0,80,20)), col="darkgray")
```

`h=(seq` tells R which axis values we want a sequence of horizontal lines at, here from 0 to 80 with an interval of 2, and from 0 to 80 with an interval of 20. Of course, should you choose to include any grid lines at all, you may wish to include only the darker major grid lines. In deciding what, if any, grid lines to use, think about what the key finding of your data is and whether lines would actually help convey that or just make the plot look messy. Here we show you how to do both the major and minor sets of horizontal grid lines, but you can always leave out any code you wish to omit (and see section 7.3.4 for further line customization options).

Step 3: Looking at the grid lines on the stripped-down plot, we can see that it might also be useful for us to make the x-axis stand out a bit more. To do this, we can simply use **abline** again, this time stating only one value where we want a horizontal line to be drawn (0, our base) and setting the colour to black in order to match the y-axis:

```
abline(h=0,col="black")
```

Step 4: Now we need to tell R that we want to overlay the existing plot of grid lines we have produced with a new plot. We prepare R to do this using the short line of code:

```
par(new=TRUE)
```

In RStudio, this line of code tells R to add the next subsequent plotting information run through the Console on top of the plot currently in the 'Plots' tab of the bottom right-hand window (rather than starting a whole new plot). A lot of graphical parameters are available to control the way figures are displayed in R, and there are therefore many arguments associated with the code **par**. We will see more of **par** later in this chapter (and see section 7.2 for more details), but remember it simply refers to the various ways you can alter the display of graphs—their 'parameters'.

Step 5: We can now draw a high-quality bar chart (with colours) on top of the current plot of grid lines with the code below:

```
barplot(abundance, names.arg=c("Field mouse","Common shrew",
"Field vole", "Bank vole","Pygmy shrew"), ylab= "Abundance",
xlab="Species",ylim=c(0,80),col="limegreen")
```

`col` is the argument we use to bring colour to our figure. Here we have opted for 'limegreen', in part because it is eye-catching, but a shade of green also seems intuitive when we are thinking about species abundance along hedgerows. The resulting bar chart (reproduced as Figure 3.3) is clearly labelled and intuitively arranged, with the added grid lines making precise values easier to read and enabling clearer discrimination of similar values.

Step 6: Any figure you produce should be accompanied by a figure caption or a figure title to explain what it shows. In scientific writing, figures are typically accompanied by a caption (see section 1.2 and Scientific Approach 1.2 for advice on writing these), but there are some occasions where you might want to give your figure a title (e.g. producing figures for posters or presentations). If you want to add a title to your bar chart, you just need to include the argument **main** in your code before running it through the Console, as below:

```
barplot(abundance,ylim=c(0,80))
abline(h=(seq(0,80,2)), col="lightgray")
abline(h=(seq(0,80,20)), col="darkgray")
abline(h=0,col="black")
```

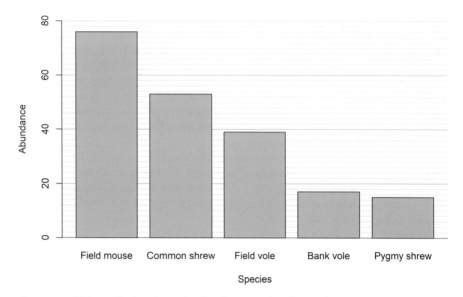

Figure 3.3 A high-quality bar chart showing the same data from Table 3.1 as Figures 3.1 and 3.2, but the use of colour makes the figure more visually appealing and the grid lines make it easier to read precise values.

```
par(new=TRUE)
barplot(abundance,main="Small mammal hedgerow survey",
  font.main=4, col.main="forestgreen",
  names.arg=c("Field mouse","Common shrew", "Field vole",
  "Bank vole","Pygmy shrew"), ylab= "Abundance",xlab=
  "Species",ylim=c(0,80),col="limegreen")
```

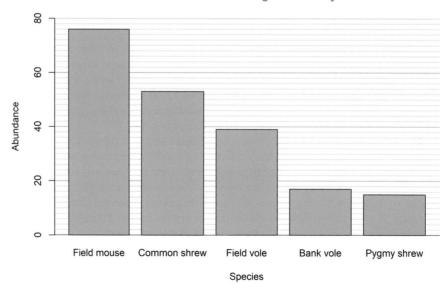

Figure 3.4 As Figure 3.3 but with a customized title added.

Notice that in the code for Figure 3.4, we not only add a title (using `main`), but we also customize this title using additional arguments that include '`.main`'. `col.main` controls the colour of the title, while `font.main` tells R we want to change the font of the title (in this case to emboldened italics, 4)—see section 7.3.2 for how to customize fonts further. However, such customization does not really add to our bar chart's message, and in fact may instead distract viewers from the data itself.

Be careful in adding any additional components to figures that could constitute unhelpful 'chart junk' (see section 1.7 for a further discussion of this).

 Key point

It is easy to produce a polished bar chart in R. Refinements don't just add visual appeal, they can also enhance ease of assimilation for your reader.

3.3   Ordinal data (ordered categories)

A study collecting data on the browsing habits of customers in a bookshop estimated the age and sex of people that spent over three minutes browsing books in the science section, and found that 51 per cent of these science section browsers were male. The percentages (across customers of both sexes) falling into different age categories are displayed for males only in Table 3.2. Because males make up 51 per cent of the sample, these percentages add up to 51.

3.3.1   Refined bar chart

Let's first rustle up a bar chart using many of the same tricks we learned earlier in the chapter.

Step 1: As this is an example of ordinal data (the categories are ordered logically by increasing age), it is not appropriate to order the bars by value as we did with the nominal data in section 3.2 (see section 1.3.2 and the online resource 'Data types' from chapter 1 for a more in-depth reminder of why this is so). This means we can keep the values in their current order when we input the data to R in a list called '`males`':

```
males<-c(6,15,12,11,4,3)
```

Table 3.2 Age structure of male browsers in the science section of a bookshop. Percentages refer to the entire sample of males and females.

Age	% customers
7–14	6
15–24	15
25–34	12
35–44	11
45–54	4
55+	3

Step 2: Now the data is in, we can lay down some grid lines for our bar chart, starting as we did before by producing a simple plot with a defined y-axis and adding our dark and light horizontal lines using **abline**. We can see from the data (reproduced in Table 3.2) that the highest recorded percentage of any male age category is 15, so a nice round number a bit above that for the y-axis to run to might be 20 per cent. Given a range of 20, having major grid lines at every 5 per cent and minor grid lines at every 1 per cent should make the heights of the bars easily readable. We will also draw on a clear x-axis:

```
barplot(males, ylim=c(0,20))
abline(h=(seq(0,20,1)), col="lightgray")
abline(h=(seq(0,20,5)), col="darkgray")
abline(h=0,col="black")
```

Step 3: Now we can overlay these grid lines with a neat and colourful bar chart again. Try the code:

```
par(new=TRUE)
barplot(males,
  names.arg=c("7-14", "15-24", "25-34","35-44",
  "45-54","55+"),
  ylab="% science section browsers", xlab="Age (years)",
  ylim=c(0,20),col="darkorchid")
```

Here we have given a description to our y-axis (using **ylab**) and our x-axis (using **xlab**), as well as defining the y-axis bounds with **ylim**. As in the previous example, we have produced a clear bar chart with very little code (see this reproduced as Figure 3.5).

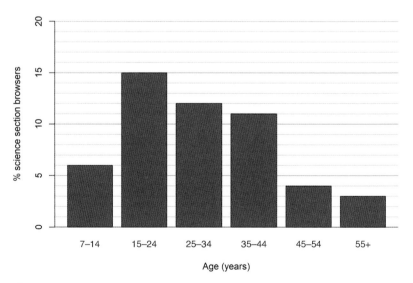

Figure 3.5 A polished bar chart showing the age structure of male browsers in the science section of a bookshop. Percentages refer to the entire sample of males and females (data from Table 3.2).

3.3.2 Further upgrade: adding minor tick marks

So far with this ordinal data we have only used code already seen with the nominal data example, but here is a new trick. We're now going to add some minor tick marks to our y-axis at every 1 per cent to further help readers interpret the values displayed.

Step 1: Produce a bar chart (like Figure 3.5) as above.

Step 2: We can add minor tick marks to the y-axis by using the package 'Hmisc' (Harrell et al. 2021) (see the guide 'R Basics' in the online resources from chapter 1 for a full explanation of how to install and activate packages in R):

```
install.packages("Hmisc")
```

Step 3: Now we activate the **Hmisc** package for our current R session using **library** and use its **minor.tick** function with the following lines of code:

```
library(Hmisc)
minor.tick(ny=5, nx=0, tick.ratio=0.5)
```

ny is the number of minor tick marks to place between y-axis major tick marks. **nx** does the same for the x-axis, though here no x-axis ticks are necessary as values are read only from the y-axis. **tick.ratio** is the size of the minor tick mark relative to the major tick mark: here, the minor tick marks are half the size of the major ones. And we can see that those minor ticks have now been added to the y-axis of our clear bar chart (we show this in Figure 3.6).

> 💡 **Key point**
>
> A bar chart might be a better choice than a table if you want to show viewers a trend or distribution in relative values across levels of your qualitative variable.

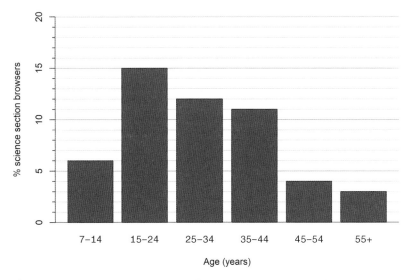

Figure 3.6 As Figure 3.5 but with minor tick marks added to the y-axis.

3.4   Grouped bar charts (comparing data from multiple samples)

Let's now compare the percentages of males of different age categories from the study in the previous example (section 3.3) with those of the females in the same study.

3.4.1 Simple grouped bar chart

Here our data is still ordinal (arranged logically by ascending age category; see Figures 3.5 and 3.6) but we now have multiple samples. Such data could not be presented in a single pie chart. However, a grouped bar chart is ideal for presenting this type of data. Grouped bar charts are useful for comparing the data between different subgroups of the main categories; in our case the subgroups are the two sexes, and our main category is age.

Step 1: First, we create a couple of lists of the values for both groups' age category percentages. The creating of lists is nothing we haven't seen before, but here we have two lists to give separate names to:

```
males<-c(6,15,12,11,4,3)
females<-c(3,12,9,13,7,5)
```

Step 2: Here comes the first new bit of code we need to show these two data sets side by side. We need to join these two lists together into a data set using the **rbind** function (this **bind**s two lists together, making each list a separate **row**):

```
browsers <- rbind(males,females)
```

If we now look at our data '**browsers**' by running its name alone through the Console, R will give us an output:

```
browsers
        [,1] [,2] [,3] [,4] [,5] [,6]
males    6    15   12   11   4    3
females  3    12   9    13   7    5
```

We can see from this how R has 'bound' together the two rows of data into the one data set. There is also a function **cbind** which binds together columns in a similar way, which we will see later in the book.

Step 3: Using this new combined data, we can now produce a simple grouped bar chart by incorporating another bit of new code into our previous bar chart code: **beside=TRUE**. We also need two colours listed for the **col** argument to

Table 3.3 The percentages of customers (separated by age class, and sex) who browsed books in the science section of a bookshop for over three minutes.

Age	% males	% females
7–14	6	3
15–24	15	12
25–34	12	9
35–44	11	13
45–54	4	7
55+	3	5

allow us to differentiate the male and female data easily (see section 1.6 for advice on choosing colours for effective and accessible figures).

Try running the following code:

```
barplot(browsers, beside=TRUE,
  names.arg=c("7-14", "15-24", "25-34","35-44",
  "45-54","55+"),
  ylab="% science section browsers", xlab="Age (years)",
  ylim=c(0,20), col=c("darkorchid","gold1"))
```

beside = TRUE is essential to create grouped bar charts—it tells R we want the corresponding bars from both males and females to be arranged beside one another. If this code gets missed out, R will produce a stacked bar chart instead; in Scientific Approach 3.1 we discuss the problem with stacked bar charts. Note that we have also defined our y-axis limits with ylim again—from looking at the data values, where 15 remains the greatest percentage across both samples, 20 still seems a reasonable maximum to set.

Scientific Approach 3.1
The problem with stacked bar charts

Different scientists have different approaches to how to present bar charts. We believe that stacked bar charts have some serious drawbacks, and we would like to explain our viewpoint here.

The key bit of code to produce grouped bar charts is **beside=TRUE**. If you miss this out, R will produce a stacked bar chart. While you may see stacked bar charts a lot, we think they are usually worse at presenting information in

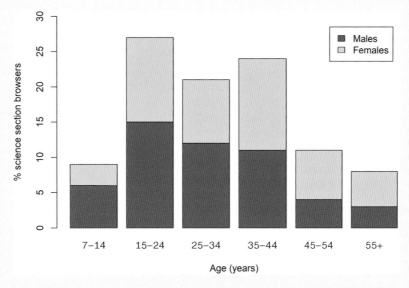

SA 3.1 Figure A A stacked bar chart showing the percentage of male and female customers of different ages who browsed books in the science section of a bookshop for over three minutes (Table 3.3).

a clear, readable way. Take the stacked bar chart example shown in SA 3.1 Figure A: it is perhaps possible to suggest that a higher percentage of 15–24-year-old males were seen browsing the science section than 15–24-year-old females, but comparing the females' percentages for each of the age groups is difficult because they start at different heights. It is also a lot more difficult to read the exact female values from the y-axis as they don't start at 0. Generally, stacked bar charts are difficult to interpret because we need a common baseline in order to judge changes in length of bar. Grouped bar charts, on the other hand, are much more effective because they avoid this issue and allow for easy comparison. A case where stacked bar charts may actually be useful for data presentation, though, is if you are presenting percentage data for subgroups that will always add up to 100 per cent when bars are stacked. See the first data set in section 3.8 and the online resource 'Authors' attempts for chapter 3' for an example of this.

Step 4: As we specified two colours in `col=` to help readers differentiate males and females, we'll also need a `legend` to explain this to the reader. You may recall this code from chapter 2, but essentially we tell R where we want the legend positioned in the quotation marks " ", what the names of the subgroups are in the `legend` argument (within the `legend` function), and what colours we have used to represent those subgroups in the `fill` argument (notice: the `legend` function uses the argument `fill` rather than `col`):

```
legend("topright",inset=0.05,legend=c("Males","Females"),
  fill=c("darkorchid","gold1"))
```

As in chapter 2, the colours are listed in the legend in the same order they were listed for the two groups in the plotting of the bar chart. Different from the pie charts of chapter 2, we are positioning our legend in the '`topright`' this time (see section 7.4 for further legend positioning and customization options) and want it a little `inset` (pulled in from the very edges of the plot a touch, so that we can see the box around it clearly). This gives us the simple grouped bar chart shown in Figure 3.7, which clearly differentiates the two samples being compared.

3.4.2 Refined grouped bar chart: adding grid lines, minor ticks, and overlying legend

Step 1: Using much the same code seen in previous examples, we can produce a neater and easier-to-read grouped bar chart if we add some appropriate horizontal grid lines and draw on a clear x-axis. We start by running this code from an R script through the Console:

```
barplot(browsers, beside=TRUE, ylim=c(0,20))
abline(h=(seq(0,20,1)), col="lightgray")
abline(h=(seq(0,20,5)), col="darkgray")
abline(h=0,col="black")
```

Step 2: Provided you are working from the same version of R on the same device as earlier, this time when adding minor tick marks (at every 1 per cent) we don't need to install or activate the `Hmisc` package again. We can just add the minor ticks straight away, using the function:

```
minor.tick(ny=5, nx=0, tick.ratio=0.5)
```

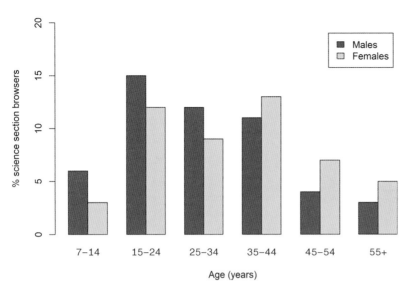

Figure 3.7 A simple grouped bar chart showing the percentage of male and female customers of different ages who browsed books in the science section of a bookshop for over three minutes (Table 3.3).

Step 3: Now we can overlay the horizontal grid lines with our bar chart again. Remember, we first need to tell R that we want to overlay this existing plot of grid lines with a new plot, with **par(new=TRUE)**, otherwise the plot in the 'Plots' tab of the bottom-right window of RStudio will be replaced rather than added to:

```
par(new=TRUE)
barplot(browsers, beside=TRUE,
  names.arg=c("7-14", "15-24", "25-34","35-44",
  "45-54","55+"),
  ylab="% science section browsers", xlab="Age (years)",
  ylim=c(0,20), col=c("darkorchid","gold1"))
```

Step 4: As this is a grouped bar chart, a legend is still necessary. When adding a legend on top of grid lines, we have to specify that we want a white background to the box so that the grid lines don't show through. This is easily done with **bg**:

```
legend("topright",inset=0.05,legend=c("Males","Females"),
  fill=c("darkorchid","gold1"),bg="white")
```

These tweaks allow us to produce the polished and clear grouped bar chart in Figure 3.8.

 Key point

If you have several samples of similar qualitative data that you want to compare, then a grouped bar chart can be particularly effective.

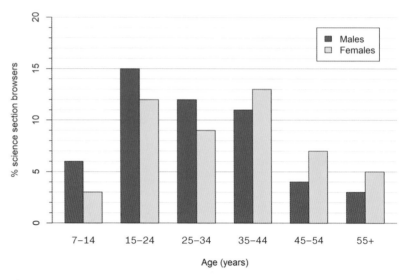

Figure 3.8 A high-quality grouped bar chart, as Figure 3.7 but with grid lines and minor ticks for readability.

3.5   Horizontal bar charts

Vertical bar charts should suit the vast majority of data sets. They are easy to interpret, and make comparisons between groups simple. However, there are a handful of occasions where you might want to present data in a horizontal bar chart. Horizontal bar charts are useful:

- when the different categories have long titles that would be difficult to include below a vertical bar
- when there are a large number of different categories that wouldn't fit across a page (e.g. 12 or more)
- when the metric displayed is duration; some people find that time is intuitively easier to interpret reading from side to side instead of up and down.

In R, it is easy to produce a horizontal bar chart by adding the argument `horiz=TRUE` to your basic bar chart code. The key differences are that all references to the x- or y-axes have to be reversed and the y-axis title might need to be added separately. Give it a go with the example below.

We are going to be looking at woodland area by forest type in Great Britain.

3.5.1   Refined horizontal bar chart

Step 1: We'll start, as ever, by inputting our data. Here the percentages covered by the different forest types (nominal data, as explained in section 1.3) are already in descending order:

```
woods<-c(49,32.1,8.1,7.9,1.8,0.5,0.2)
```

Step 2: First, in some cases with horizontal bar charts we might need to add a bit more space in the outer margin area to the left, to accommodate longer

Table 3.4 The percentage of total area different forest types covered of woodland areas in Great Britain from a survey carried out between 1990 and 2003 (Forestry Commission 2003).

Forest Type	% total area
Conifer	49.0
Broadleaved	32.1
Open space	8.1
Mixed	7.9
Felled	1.8
Coppiced	0.5
Windblow	0.2

y-axis labels—this is certainly true of our woodland data set if we are to use the full names of forest types. This is done with **par** (see section 7.2 for more details on the use of **par**). **par** can be used to change the size of margins and **oma** refers to the 'outer margin area' specifically, with the listed numbers corresponding with space at the bottom, left, top, and right respectively. By default, **par(oma = c(0,0,0,0))**. Here we tell R we want the left outer margin wider:

```
par(oma = c(0, 6, 0, 0))
```

If your y-axis labels are going to be short, though, there is no need to create the extra space and your y-axis title can be included in your main **barplot** code simply as **ylab** (as in this chapter's previous examples).

  Step 3: Next we set out axis limits and add grid lines using very familiar code, but this time we include the all-important **horiz=TRUE**, set our x-axis limits rather than the y-axis limits using **xlim**, and make sure we add vertical (v) rather than horizontal (h) lines, as the x-axis is where values will be read from this time. Note that the axis we are drawing in black is also vertical (v) as it is the y-axis rather than the x-axis that we are adding:

```
barplot(woods, horiz=TRUE, xlim=c(0,50))
abline(v=(seq(0,50,2.5)), col="lightgray")
abline(v=(seq(0,50,10)), col="darkgray")
abline(v=0,col="black")
```

Step 4: We then add minor tick marks, using the package **Hmisc** again (see the guide 'R Basics' in the online resources from chapter 1 for a recap of installing and activating packages), but this time adding them to **nx** not **ny**:

```
library(Hmisc)
minor.tick(nx=2, ny=0, tick.ratio=0.5)
```

Step 5: Now we draw the bar chart over the top (starting with **par(new=TRUE)** as we have seen before), including **horiz=TRUE** and, new to this example, **las=1**. **las** represents the style of axis labels relative to the axes whereby: **0**=parallel, **1**=all horizontal, **2**=all perpendicular to axis, **3**=all vertical. Note that we again include the same x-axis limits (**xlim**) as we did when producing the grid lines. We also miss out the y-axis title (**ylab**) from our code in this case, as this needs to be made horizontal (and positioned so that it does not overlap our long y-axis labels) separately:

```
par(new=TRUE)
barplot(woods, horiz=TRUE, xlim=c(0,50),
names.arg=c("Conifer","Broadleaved","Openspace","Mixed",
"Felled","Coppiced", "Windblow"), xlab="% of total area",
las=1, col="forestgreen")
```

We can see that our y-axis labels now fit neatly next to our bars, thanks to the space we created for them in the earlier section of this example. We have also chosen an intuitive and fitting colour ('**forestgreen**') for this data.

Step 6: As mentioned above, if your y-axis labels are going to be short, then there is no need to create the extra space, and your y-axis title can be included in your main **barplot** code as **ylab**. Here, though, we now need to add our y-axis title using **mtext**.

In the following code, **side** is simply which side of the plot (1=bottom, 2=left, 3=top, 4=right) we want the text added, **line** refers to which margin line (an R parameter) we want the text, starting at 0 counting outwards; 6 works well here (see sections 7.2 and 7.3.5 for more details on space, margins and **mtext**). The \n code interrupting the y-axis title tells R to start a new line here to space the title out neatly:

```
mtext("Forest\n type", side = 2, line = 7,las=1)
```

Putting all this together, the effective horizontal bar chart shown Figure 3.9 should pop up in the 'Plots' tab of the bottom-right window in RStudio. You may need to stretch out the RStudio window containing the plot to see the y-axis labels more clearly (see the 'R Basics' guide in the online resources from chapter 1 for details on using RStudio).

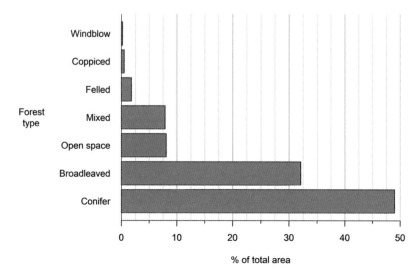

Figure 3.9 A high-quality horizontal bar chart, displaying the percentage of total area different forest types cover of woodland areas in Great Britain (Forestry Commission 2003), from Table 3.4.

Step 7: Remember now to reset the outer margins for your next R graphs:

```
par(oma = c(0,0,0,0))
```

 Key point

Horizontal bar charts can be useful if you have a lot of categories to your data, those categories have long names, and/or if you want to display durations.

3.6   Tips and tricks

Just to show off what R can do, let's customize the bars of a bar chart a bit more. To be clear, we do not recommend customizing any figure to the extent we are about to do, but we use this as an example of the available commands. We'll return to our ordinal data set on the ages of customers browsing the science section of a bookshop, this time focussing on just the females data from section 3.4.

Step 1: Input the data in the same way we did in section 3.4, with the code:

```
females<-c(3,12,9,13,7,5)
```

To customize your bars, the following bits of code all have to be included in the commands where you plot the chart, but we'll have a look through them individually first. To begin with, the width of bar borders can be changed by preceding the bar chart code with:

```
par(lwd=)
```

This defines the line widths as whatever number you choose (see section 7.2.4 for more details on establishing graphing conditions ahead of plotting, and section 7.3.4 for further line customization). Here we'll try thick bar borders with:

```
par(lwd=3)
```

We can also add shading to bars using:

```
density =, angle =
```

You may recognize these arguments from the online resource 'Shading and 3D' from chapter 2. The **density** part refers to the density of shading lines, in lines per inch, while **angle** gives the slope of shading lines, given as an angle from horizontal in degrees (counterclockwise). We can define these to apply to all bars by assigning each only one value (e.g. **density=10, angle =75**) or we could make the shading different for each bar using **c()** to list different densities for each bar. Here we'll shade bars with increasing density:

```
density = c(5,10,15,20,25,30), angle = 75
```

The colour of your bar outlines can also be customized with:

```
border =
```

We can either have this the same across all bars (e.g. **border = "red"**) or have a different colour for each bar (e.g. **border = c("red","black","brown", "magenta","lightblue","seagreen")**). Alternatively, we can make borders the same colour as the shading lines using:

```
border = TRUE
```

Step 2: Now try this full bit of code, with all the extra customization detailed above:

```
barplot(females,col="white",ylim=c(0,20))
abline(h=(seq(0,20,1)), col="lightgray")
abline(h=(seq(0,20,5)), col="darkgray")
par(lwd=3)
par(new=TRUE)
barplot(females, main="Female customers", font.main=2,
  col.main="darkblue", names.arg=c("7-14","15-24",
  "25-34","35-44","45-54","55+"),
  ylab="% science section browsers", xlab="Age (years)",
  ylim=c(0,20), density = c(5,10,15,20,25,30), angle = 75,
  col=c("red","orange","yellow","green","lightblue","violet"),
  border = TRUE)
```

Note that when plotting the chart to establish grid lines we have this time made the bars `col= "white"` so that the shading will be clear when we add our customized chart on top. We have also added a title this time (using **main**), made it bold (**font.main=2**—see section 7.3.2 for further font customization), and changed its colour from black (using **col.main**). Remember, though, that most scientific figures should be accompanied by a detailed caption instead of

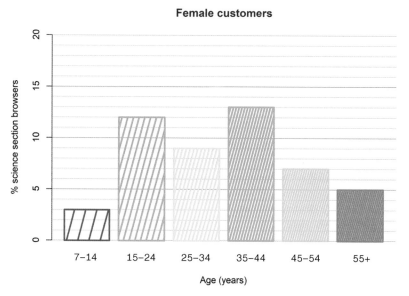

Figure 3.10 A bar chart with several unnecessary customizations showing the percentage of female customers of different ages who browsed books in the science section of a bookshop for over three minutes (Table 3.3).

a title (see section 1.2 and Scientific Approach 1.2 for advice on writing figure captions).

The bar chart in Figure 3.10 should appear in the 'Plots' tab of the bottom right-hand window in RStudio. Honestly, we do not think that this looks as good as solid bars, and would advise against using multiple colours unless they serve a purpose (see section 1.6 for further discussion of colour choice), but if you disagree now you know how to do this. More broadly speaking, we would again advise caution when customizing figures—any component included should enhance, rather than distract from, the figure's key message (see section 1.7 for advice on chart design and a discussion of 'chart junk').

Step 3: Remember to set **par** line widths back to default (which is size 1) afterwards so that R doesn't draw everything else in your session with thick lines!

```
par(lwd=1)
```

 Key point

R is very flexible, so you can be very creative and find a way to make R produce a figure exactly the way you want it.

3.7   Non-zero lowest y-axis values on bar charts

While R will by default assume that the y-axis of a bar chart starts at 0, this does not have to be the case. Figures in reports, popular media, and in scientific papers do not always have their y-axis set with 0 as the minimum. Some people argue that this is misleading, and that figures should always have the y-axis set to begin at 0. If this is not the case, the data could be considered misrepresented, as it skews readers' perception of what is being shown. Values presented in bar charts in particular are intuitively compared assuming a base of 0. Thus, where this is not the case, differences between data points can seem exaggerated and more striking. Hence this technique could be used to mislead people, and there are many instances where those producing such figures have been accused of bias or trying to push an agenda on the basis of having a non-zero start to the y-axis.

However, we feel that sometimes starting the y-axis at 0 can make patterns in the data harder to see, and can also mean that much of the space taken up by the figure has low informational content. Imagine a situation where you are comparing only very large values. The slight differences in values between data measurements might have significant consequences for whatever those values represent. But if you plot all these values from 0 it could be really difficult to even notice small differences between the large values. Sometimes, we want to use figures to convey information and make a point—forcing the y-axis to start at zero can, in some cases, obscure and confuse this point.

To us, the simplest and most honest solution when a non-zero base seems attractive but you don't want to risk misleading your reader, is to include both variations of the same graph. This way, any important patterns in your data can be highlighted, but the data is still presented in context. We provide an example of this in Scientific Approach 3.2.

Scientific Approach 3.2

An effective approach to presenting figures with y-axes at non-zero values without misleading

One way you could present both a figure that started its y-axis at zero and a figure that started at a non-zero value would be simply to plot two figures next to each other using **par(mfrow=c(1,2))** or **par(mfrow=c(2,1))** (see section 7.2 for details on **par**). As a neater alternative, though, we recommend producing a 'zoomed in' version of the figure with the altered non-zero y-axis and then insetting a second figure of the whole picture (with y-axis starting at 0) to give 'the bigger picture'. You can think of this like Russian dolls—a plot within a plot.

SA 3.2 Figure A demonstrates how an inset can be used to present both a 'zoomed in' and complete picture of a data set. The main bar chart demonstrates how Canada's per capita greenhouse gas emissions have decreased, most notably between 2007 and 2009, but also seemingly stepping down again between 2015 and 2017 (Environment and Climate Change Canada 2020). However, looking at the non-zero y-axis, the decrease only spans around 3 metric tonnes of CO_2 equivalent per capita, which is a small fraction of total emissions. The insetted bar chart shows us what this decrease looks like against a y-axis that starts at zero. The inset here certainly makes the pattern of the data look less dramatic,

but if such changes in emissions are significant and herald important environmental benefits then it is well worth highlighting these with a non-zero y-axis. Thus, the two charts together show us clearly that Canada has reduced emissions in recent years; but they also highlight that these reductions are a modest fraction of overall emissions. Presenting both versions of the graph is a little more work, but we think it really helps the reader to appreciate all aspects of this data. As we will show you in chapter 8, the extra work involved in producing such insets can be very modest.

If you are interested in using insets to provide context while highlighting important patterns with non-zero axes in your figures, see section 8.2 for a full example using the same data as SA 3.2 Figure A.

Also notice that in this example we use a bar chart to present **quantitative data**, unlike the rest of this chapter. As we will argue in chapter 5, this is not an approach that we strongly favour. However, you will see bar charts commonly used this way, so we thought we would introduce this to you now and discuss it more in chapter 5 and in its associated online resource 'Bar charts for quantitative multiple-samples data'.

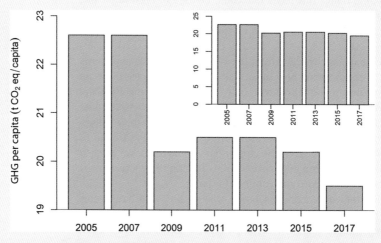

SA 3.2 Figure A Biennial Canadian per capita greenhouse gas (GHG) emissions from 2005 to 2017 (Environment and Climate Change Canada 2020).

Key point

It seems very natural to start your y-axis at zero, but sometimes this might not be the ideal choice. In such cases, you need to be careful not to unintentionally mislead your readers when starting from a non-zero value.

3.8   Stretch your understanding

To consolidate the ideas that we have introduced in this chapter, we recommend you have a go at producing bar charts of the data sets we provide for chapter 2. Not only will this hone your coding skills in R, but it will allow you to compare for yourself the advantages and disadvantages of the bar charts you produce compared to the pie charts and tables provided in chapter 2. For further practice, have a go at making the best bar charts you can of the following two data sets.

The first data set is based on a study by Inari et al. (2005) that aimed to explore the potential impact on native bumblebees of introduced *Bombus terrestris* bumblebees in Hokkaido, Japan. Table 3.5 shows the percentage of different castes of *B. terrestris* individuals that were trapped across three separate months. By visualizing this data as a bar chart, the changes in caste structure over these months may be easier for viewers to interpret. The data collected in each month is qualitative (based on counts of trapped individuals) but ordinal, as there is a logical sequence to June, July, and August. While we would normally recommend grouped bar charts for cases such as this, where we are comparing the different samples (castes), you may instead choose to use a stacked bar chart here. This is, firstly, because the data is in percentage form and adds up to 100 per cent for each month (see Scientific Approach 3.1 for more details on stacked bar charts) and, secondly, because we are interested in the overall trend of changing caste proportions—reading precise values for each caste is not essential for this.

The second data set comes from a report by the Alzheimer's Association (2020), showing how the number of deaths from Alzheimer's disease in the United States more than doubled between 2000 and 2018. In Table 3.6, this is shown in relation to the percentage change in deaths recorded due to a number of other selected conditions including heart disease, the leading cause of death in the US and worldwide (which actually decreased over the same time period). This data set is nominal (there is no logical ordering to the conditions) and includes some negative values. But, do not fear! Bar charts are good at showing

Table 3.5 Monthly changes in the percentage of trapped *Bombus terrestris* individuals of different castes (numbers based on a study by Inari et al. [2005]).

	% trapped individuals		
Caste	**June**	**July**	**August**
Queen	59.6	5.4	8.7
Worker	40.4	94.6	77.4
Male	0	0	13.9

Table 3.6 Percentage changes in selected causes of death in the United States between 2000 and 2018. Based on a figure by the Alzheimer's Association (2020) created from data from the US National Centre for Health Statistics (CDC WONDER n.d.; Tejada-Vera 2013).

Cause of death	% change
Breast cancer	1.5
Prostate cancer	1.3
Heart disease	−7.8
Stroke	−11.8
HIV	−62.5
Alzheimer's disease	146.2

values less than zero too. You just have to have a think about where you want your y-axis to run from and whether it might be good to have a darker and/or thicker graph line at zero.

Remember, if you look at the online resources for this book, you will find our individual attempts at these exercises, including the code we used ('R script for Authors' attempts chapter 3') and the finished charts ('Authors' attempts for chapter 3'). You will also find our comments on why we made the design decisions we did, and what we think of the other author's attempt.

Chapter Summary

- Bar charts are an effective way to display both nominal and ordinal data, but the different data types may influence your design choices.
- Using R, the basic code for bar charts is: `barplot(x, names.arg=c())` but there are many ways bar charts can be customized to improve their clarity and visual appeal.
- Bar charts are more effective than pie charts because: i) their linear arrangement of bars reduces misinterpretation of relative values, ii) low-valued categories are as visible along the x-axis as high-valued categories, iii) there is more space to label each category, and iv) comparison across data sets is easier than across several pie charts.
- A bar chart might be a better choice than a table if you want to show viewers a trend or distribution in relative values across groups.
- A table may be a better choice than a bar chart if you want the reader to appreciate the absolute values in the data.
- Grouped bar charts are ideal for presenting qualitative data on the same categories from across multiple samples.

- Data may be most effectively presented as a horizontal bar chart when: i) the different categories have long titles, ii) there are a large number of different categories that wouldn't fit across a page (e.g. 12 or more), or iii) the metric displayed is duration.
- There are advantages and disadvantages to starting bar chart y-axes at a non-zero value; sometimes using an inset can allow you to have the best of both worlds.

Online Resources

The following online resources are available for this chapter at www.oup.com/he/humphreys-obp1e:

- R script for chapter 3
- Authors' attempts for chapter 3
- R script for Authors' attempts chapter 3

Further Reading

- 'A Complete Guide to Bar Charts':
  https://chartio.com/learn/charts/bar-chart-complete-guide/
- 'Bar Charts Should Always Start at Zero. But What about Line Charts?':
  http://www.chadskelton.com/2018/06/bar-charts-should-always-start-at-zero.html
- 'Non-zero Baselines: The Good, the Bad and the Ugly':
  https://thenode.biologists.com/non-zero-baselines-the-good-the-bad-and-the-ugly/resources/
- 'Does the Axis Have to Start at Zero? (Part 2—Bar Charts)':
  https://digitalblog.ons.gov.uk/2016/07/23/does-the-axis-have-to-start-at-zero-part-2-bar-charts/
- 'It's OK Not to Start Your y-axis at Zero':
  https://qz.com/418083/its-ok-not-to-start-your-y-axis-at-zero/

4 PRESENTING SINGLE-SAMPLE DATA: HISTOGRAMS AND BOXPLOTS

Learning objectives

By the end of this chapter you should be able to:

- Explain when histograms and boxplots are good data presentation options, and what the advantages and disadvantages of each figure type are relative to the other.
- Produce clear and effective histograms from single-sample data.
- Produce clear and effective boxplots from single-sample data.
- Understand how the 'whiskers' on a boxplot are interpreted.
- Customize histograms by changing such features as bin size, cell colour, and axis features.
- Customize boxplots by adding text labels, changing box colour, and editing axis features.

4.1 Introduction: when do I have single-sample data and what can I do with it?

Here we will be looking at two main graphing types we can use to present data from a single sample: histograms and boxplots. For both figure types, the data we present will come from one particular sample where for each individual within the sample we have one single measurement in our data set. This measurement should be continuous, or be discrete but have quite a lot of different values. So, this might be the masses or heights of a sample of first-year university undergraduates. Something like the total value of the coins each undergraduate has in their possession could also be handled this way (because you would probably get a wide spread of discrete values across the sample). Histograms and boxplots are both suitable for presenting this sort of data, but which choice is most suitable depends in part on what you intend to communicate from the data (as we will discuss later). If the single sample you are looking at contains only a few different values (as is likely if the quantity measured on each undergraduate is, for example, number of siblings, or number of university society memberships) then this data is best displayed using one of the methods covered in chapters 2 and 3. If you feel that a rule of thumb for 'only a few different values' would help you, then 'seven or fewer' generally works for us.

Histograms are an excellent way of visually displaying the distribution of numerical data; it is always valuable to visualize this prior to performing any statistics on data. But a bit of thought needs to go into how best to group or 'bin' your data in preparing the histogram. We will demonstrate design features and histogram-specific rules of thumb that can greatly improve their look and interpretability, but the basic code in R is simply `hist(x)`. Like histograms, boxplots can also give a good sense of the spread of data, but additionally they make some key descriptive statistics easy to interpret and allow for easy identification of outliers.

We will show you some effective ways to refine boxplots, using some customization code already covered in chapters 2 and 3 as well as some new additions, but again the basic code in R is easy: `boxplot(x)`. In this chapter, we will show you how to produce both figure types from the same single-sample data set to allow you to see the differences for yourself and judge which is best for your purposes.

 Key point

Histograms and boxplots can be good options for displaying a single sample of **quantitative data**.

4.2   An example of single-sample data

Before we start producing any figures, we need to have a look at our data. Unlike in previous chapters, the data set we will be using as an example is so big that it would be too laborious to manually input each value into R. Instead, it can be loaded into R from the .csv file 'extinction_data' (downloadable from the online resources), and given the name **mydata**. As a reminder of how to load data sets into R (and why we have prepared the data as a .csv file), see the 'R Basics' guide from the online resources for chapter 1.

The data set 'extinction_data' stems from a report by Urban (2015) that synthesized the findings of published studies in order to estimate how global extinction rates might accelerate with predicted climate change. Specifically, the data lists 130 studies, and for each study gives a number between 0 and 100 that is the mean predicted percentage of species at risk of extinction. Naturally, predicted extinction rates will vary between continents, with type of organism, and depending on the assumptions included in modelling work; but getting some sense of the overall global impact of climate change on species diversity is important for policymaking, conservation, and raising awareness.

Once you have loaded in the data set, have a quick look at it using the code **View(mydata)**. You will see in the first column ('**study**') the first author surname and year of publication for each study, and in the second column ('**percent**') the mean predicted percentage of species at risk of extinction. Each row provides us with a single percentage value for an individual study. Visualizing this single-sample data as a figure will help us to interpret the data more easily, and to identify any overall patterns across the 130 studies in predicted extinction rates.

For the sake of simplicity throughout the chapter, we will assign a name to the column of values we want to plot with the command:

```
values <- mydata$percent
```

It is useful to get into the habit of giving data columns convenient names when you have loaded data in from an existing spreadsheet. Here we are assigning the name 'values' to the column of data called 'percent' from the 'mydata' data set. The $ is how R identifies the different columns collected within a data set.

 Key point

It is always worth familiarizing yourself with how exactly your data is recorded, so you can get R to manipulate that data the way you want.

4.3   Histograms

4.3.1 Simple histogram

Step 1: As soon as your data is loaded into R, you can easily have a look at the distribution of predicted extinction rates across studies (as this is the variable of interest) with the code:

```
hist(values, ylab= "Frequency", xlab="Predicted extinction
rate %", main=NULL))
```

ylab tells R the y-axis label to use and xlab tells R the x-axis label. main=NULL simply tells R we do not want the figure to have a title, as the hist function includes a title by default. You could substitute in a title of your choosing here within " " quotation marks if desired but, as we discussed in section 1.2 and Scientific Approach 1.2, scientific figures are typically accompanied by a caption instead.

If our main interest in the predicted extinction rates variable is its distribution—that is, the frequency with which the different values in the data set occur relative to others—then Figure 4.1 looks like a good start. You will notice from Figure 4.1 that the x-axis of a histogram differs from bar charts in that it is not separated into different categories but, instead, 'cells' (not 'bars') stem from different *bins*. This is because histograms display continuous data, while bar charts plot categorical variables. Binning the range of values for a histogram is where you divide up the entire range of values into a series of intervals (called *bins*) and then count how many of your data values fall into each bin. Often you will need to play around with the number of bins you use (i.e. how finely you partition the range of values) in order to find the most effective way of presenting the data. We will explain some rules of thumb you may choose to follow for this next.

4.3.2 How many 'bins' should my histogram have?

If you are using R to plot a histogram, it will automatically have a go at deciding what might be a suitable number of bins for your data when you use the hist function. Often R makes a decent job of this, but it is always worth considering whether specifying a different number of bins would improve the visual appeal and interpretability of your histogram. Figure 4.2 at the end of

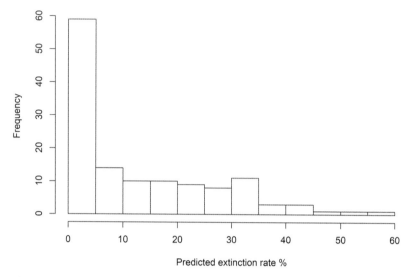

Figure 4.1 A simple histogram showing the distribution of averaged percentage predicted extinction rates across 130 published studies (Urban 2015).

section 4.3.2.2 shows several histograms differing in their bin number and sizes. If you are unsure what we mean by the term 'bin', it is essentially the x-axis range spanned by each cell.

4.3.2.1 Rules of thumb

There are lots of different rules of thumb that suggest ways to choose the right number of bins. Two of the most commonly used ones are:

Sturges' rule:

$$K = 1 + 3.322\log(n)$$

Rice rule:

$$K = 2n^{1/3}$$

For both rules, K = number of bins, n = number of observations in the sample. For Sturges' rule, log = logarithm (to the base 10) of the number.

Table 4.1 details the number of bins these two rules suggest (rounded to the nearest whole number) for a range of sample sizes. From this, we can see that

Table 4.1 The different numbers of histogram bins Sturges' rule and the Rice rule recommend for a range of sample sizes.

Sample size (n)	Sturges' rule	Rice rule
5	3	3
10	4	4
50	7	7
100	8	9
1000	11	20

they offer similar suggestions at small sample sizes but differ much more at larger sample sizes. So these two rules are probably a decent guide as to what number of bins would be good, but they do not always agree. There is no 'right' number of bins for a histogram, but there are certainly some that are less helpful. We will explore this concept further next.

4.3.2.2 Playing with bin number

To demonstrate how the number of bins you choose for your data can impact the look of a histogram, we have created a nonsense data set called 'bin_play' for you to download from the online resources as a .csv file. As a reminder of how to load existing data sets into R (and why we have prepared the data as a .csv file), see the 'R Basics' guide from the online resources for chapter 1. We also provide all the code we use to look at the 'bin_play' data in the 'R script for chapter 4' in the online resources. Once you have the data set downloaded (we called our data set 'play'), we recommend having a look at the data and giving the data to plot a convenient name so that we do not have to use the long-form name with the $ throughout (as we did in section 4.2)—we just stuck with the column name 'numbers':

```
numbers <- play$numbers
```

Step 1: Both Sturges' rule and the Rice rule depend on us knowing 'n', the sample size of our data. We can easily get R to tell us this using the length function:

```
length(numbers)
```

This tells us that we have 36 values in the data set, or n=36.

   Step 2: We can now get R to calculate the number of bins suggested by Sturges' rule, by typing:

```
1+3.322*(log10(36))
```

Our '36' has replaced 'n' in the formula from section 4.3.2.1 and the * means multiply. log10 is what R requires to determine the base logarithm of a number, which is what we need here to calculate log. Altogether, this gives us a value of 6.17.

   Step 3: Turning now to the Rice rule, we can run the following in R:

```
2*(36^(1/3))
```

Again we have replaced the 'n' in the formula from section 4.3.2.1 with our known sample size of '36'. Here ^ means 'to the power of'. We also have to be careful to put the 1/3 in brackets of its own so that R performs the steps of the calculation in the correct order. This gives us a value of 6.60.

   Step 4: As the values produced by both rules are similar, it makes sense for us to figure out an average for the two in order to determine a suitable bin number. We first list the two values we want to find the mean of and then tell R to find the mean of that list:

```
bins<-c(6.17,6.60)
mean(bins)
```

The average number of bins suggested by the Sturges' and Rice rules, therefore, is 6.385, which rounds down to 6.

   Step 5: Now, to specify this number of bins in a histogram of the data we can use the argument breaks= as in the code below:

```
hist(numbers, breaks=6, main="6 bins", ylab="Frequency")
```

Step 6: If you ran the above step in R, you will see that the resulting histogram looks pretty clear. Here we'll compare that histogram with two others—one with two bins and one with 20 bins:

```
par(mfrow=c(1,3))
hist(numbers, breaks=2, main="2 bins", xlab="Numbers",
ylab="Frequency")
hist(numbers, breaks=6, main="6 bins", xlab="Numbers",
ylab="Frequency")
hist(numbers, breaks=20, main="20 bins", xlab="Numbers",
ylab="Frequency")
par(mfrow=c(1,1))
```

The code **par(mfrow=)** is used to combine multiple plots into one figure (see section 7.2.2 for more details on how to prepare multi-panel plots). The first number designates the number of rows you want, and the second number designates the number of columns. Here we tell R at the beginning that we want the following plots in one row but three columns, and at the end we set it back to the default of plotting figures individually (one row, one column). Our three plotted histograms are shown in Figure 4.2.

The main thing to notice in Figure 4.2 is how much the choice of bin number affects the look of the histogram. Too few bins, and the shape of the data is obscured. Too many bins, and the histogram looks 'choppy', with bins that contain zero data values. We want to find the number that gives the 'nicest' shape but also has easy-to-absorb bin values with round-number end points.

Notice also that in our '2 bins' plot, R hasn't actually let us have just two bins. This is because the **breaks=** number is just a suggestion to R, and R will then calculate the best number of cells with the suggestion as a guide. R will take your initial suggestion and stick close to it but also try and make the breaks between bins occur at easy-to-digest round numbers. Another possibility is to set the '**breaks**' argument as a list of values, so that we can set a specified number of bins of specified sizes. We will use **c()** to do this in the next section, in the context: **breaks=c()**.

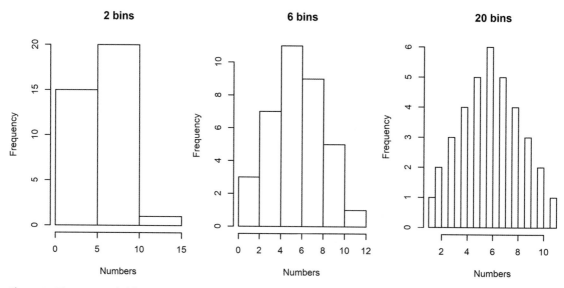

Figure 4.2 Three example histograms of the same nonsense data, demonstrating the impact of different bin numbers on the overall shape.

We have seen here how the choice of bins matters for aesthetics, but also for interpretation of your data. To summarize, it's often best to experiment with a few different bin numbers, and then pick the one that yields the most mean-ingful, useful histogram for your data. However, as we saw in Figure 4.1, if you don't specify any number of bins, then R will guess a suitable number itself– and it often guesses well.

4.3.3 Simple histogram with specified bins

Having considered how bin number can affect the look of a histogram, let us re-turn to our single-sample data set of the mean extinction rates predicted by 130 studies. If you like you can check for yourself that for this data set the Sturges' and Rice rules suggest 8 and 10 bins respectively, but R chose a slightly higher number of 12 bins in Figure 4.1 to give easy-to-digest bins that break every 5 per cent. If you happened to think that actually half as many bins that were twice as wide looked better, then the **breaks** parameter offers us two ways to do that: either by specifying the number of breaks or the position of each one. The code specifying different bins could either look like:

```
hist(values, breaks=6,ylab= "Frequency", xlab="Predicted
extinction rate %",main=NULL)
```

or:

```
hist(values, breaks=c(0,10,20,30,40,50,60),ylab= "Frequency",
xlab="Predicted extinction rate %",main=NULL)
```

Either way, using this data set the resulting revised histogram is shown in Figure 4.3. It is important to note that in the first version of the code (where we just tell R we want '6' breaks), this only produced exactly six cells in Figure 4.3 because this gives easy-to-digest bins that break at every 10 per cent. As seen in section 4.3.2 (and Figure 4.2), when given solely as a single number, the **breaks** argument is just a suggestion to R, and R will still calculate the best number of cells using this only as a guide. However, in the second version of

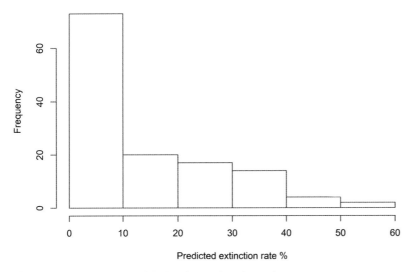

Figure 4.3 As Figure 4.1 but with the bin number changed to 6.

the code (where we specify each break in a list of values) you could force the bins to break at any values you desire and R will execute the command exactly rather than using it simply as a guide.

We think the 'shape' of the data isn't any easier to digest in this simplified histogram (shown in Figure 4.3) with only six bins, and you have of course lost some information when you aggregated into bigger bins, so we think that R made the best initial choice of 12 bins. Generally, R will make good choices for the number of bins, but here we have shown you the tools to give you any bin size and number you want. We advise caution in forcing break points between bins that R disagrees with, however, as awkward, unrounded cell ranges can make figures a lot harder to interpret.

4.3.4 Extreme values and histograms

As an aside, we see a weakness to histograms with our extinctions example. They are not good at showing you if lots of sample individuals take the same value. This can often occur at the extremes when the data is constrained in the range of values it can take (e.g. percentages must be between 0 and 100). In our case, our histograms show that many of the studies predict low extinction rates (less than 5 per cent), but it does not show us how many of these values were actually zero. In Figure 4.4 we have preprocessed our sample to remove zero values, counted those removed values, plotted only the non-zero values in a histogram with 12 bins, and provided some text in the caption to say how many zero values we have removed. One benefit to doing this is that the first bin of the histogram is no longer quite so much bigger than all the rest, so the y-axis does not have to be stretched so much to accommodate it, and so differences between the other bins can be more easily seen. You can find the code that we used to do this in the 'R script for chapter 4' in the online resources.

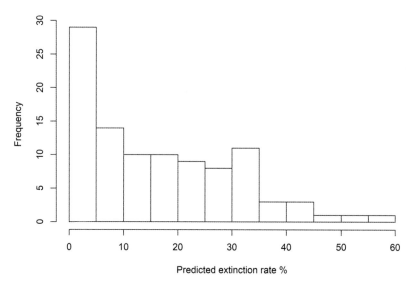

Figure 4.4 A simple histogram showing the distribution of averaged percentage predicted extinction rates across 100 of 130 published studies (Urban 2015); 30 studies predicting 0 per cent extinction rates were excluded from this figure.

In general, selecting only part of the data to display can be a good option when (as in our case) repeated instances of an extreme value is a dominant feature of the data. However, you might feel that not plotting the zero values (even though we mention them in the figure caption) could present a misleading visual impression of the data. If you did feel that this was an important issue then you could simply continue to lump zero values together with other small values as we did previously, or in section 4.3.5 we offer you another approach. You need to decide on the best approach based on what you want readers to take away from your figure. Similarly, in cases where the values you are plotting involve a very broad spread of values, you might consider changing how you communicate the data to readers by log-transforming it, and we discuss this further in Bigger Picture 4.1.

Bigger Picture 4.1
Dealing with a broad spread of values

A common way of presenting data with a broad spread of values and/or that is highly skewed is to log-transform it. This involves replacing each value with its logarithm. In R, the function **log** produces natural logs (logarithm to the base $e \approx 2.718$), whereas the function **log10** produces logarithms to the base 10. Logarithms to the base 10 are usually easier to interpret because if a value is 10^a then log10 of that value is simply a (e.g. log10(10) = 1, log10(100) = 2, log10(1000) = 3). Whichever base is used, logged data is immediately harder to interpret than raw values and so is probably best avoided unless necessary. But where your data spans several orders of magnitude, structure in the data involving low values can get lost in a histogram of raw values, and here logging can be a great option. A histogram of logged data compresses a broad spread into something easier to visualize, where the structure of low as well as high values is more effectively showcased. To demonstrate, we have had a look at a data set called 'mammals', from an R package called '**MASS**' (Venables and Ripley 2002), which includes the average body weights for 62 species of land mammals. From the untransformed data plotted as a histogram in panel **a.** of BP 4.1 Figure A, we can see that the structure of body weights <1000kg is lost entirely because of the extreme range of values the x-axis has to encompass. That is, we know that almost all our sample have masses

below 1000kg, but we cannot see any greater detail than that. Increasing the number of bins might help a little, but the essential problem will remain that most of our data will be squashed up at the left end of our figure. If we look at panel **b.** of BP 4.1 Figure A, where the same data is presented as logarithms to the base 10, we have a much clearer and more informative visualization of the distribution of different body weights across the 62 land mammals. For example, we can see that most of the sample lie between log values of -1 and 3. You can convert those values in your head, to see that most of the sample have masses between 0.1kg (10^{-1} = 0.1) and 1000kg (10^3 = 1000), and the values between 1 and 10kg are particularly common. We get much more detail because by using logged values we spread the sample out better across the figure, rather than having most of it bunched at one end. So, using logged values can be really effective when your data spans a really broad range of values. By logging to the base 10 and selecting integer end points between bins, you can minimize the challenge of interpreting logged data. Logging values ahead of plotting is trivial to do in R, and the code used to generate BP 4.1 Figure A is provided in 'R script for chapter 4' in the online resources. See section 8.5 and the online resource 'Authors' attempts for chapter 6' for more discussion and examples of plotting logged data.

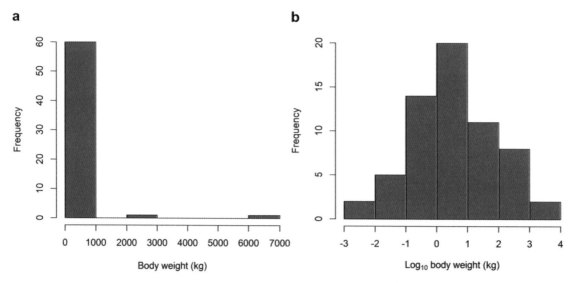

BP 4.1 Figure A Average body weights for 62 species of land mammals (Venables and Ripley 2002), plotted as: a) raw data (untransformed), and b) logarithms to the base 10 (transformed).

4.3.5 Refined histogram: customizing axes, and adding grid lines and colour

Once you have decided the best number of bins for your histogram, you can turn your attention to the aesthetics of your plot. For one, you may want to show the range of values each bin comprises, rather than individual numbers running along the x-axis. Instances where this may be particularly valuable include cases where each cell does not represent the same range of values. To demonstrate this, we will now explore another way we could deal with the 30 studies reporting 0 per cent extinction rates in our data set, rather than excluding them altogether: that is, to present them as an individual initial cell alongside our other bins spanning sensible, easy-to-digest ranges of 5 per cent.

Step 1: In order to include a bin of all the studies that report 0 per cent extinction rates, we will need to set the first value in the **breaks** argument list as a negative value (below 0 per cent). If we choose to keep 12 bins with sensible, rounded ranges spanning 5 per cent (and therefore even cell widths), then we will need our first bin to run between -5 per cent and 0. That is what we have done in the code below:

```
hist(values, breaks=c(-5,0,5,10,15,20,25,30,35,40,45,50,55,60),
  ylab= "Frequency",
  xlab="Predicted extinction rate %",main=NULL)
```

We could also have written out the list of bin **breaks** more concisely as `c(seq(-5,60,5))`, which tells R that the list of numbers is a **sequence** running between –5 and **60** with intervals of 5. This code gives us the histogram in Figure 4.5.

Figure 4.5 now includes our cell showing only studies reporting 0 per cent extinction rates. However, this cell is messily situated to the left of the 0 on the x-axis, where there is no x-axis at present. It would also be easier to interpret

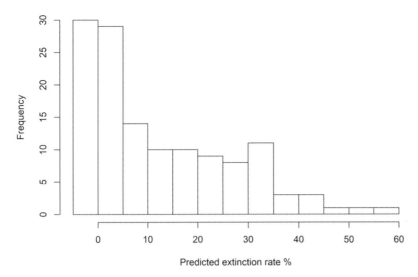

Figure 4.5 As Figure 4.4 but with the bins specified to include a separate cell for the 30 studies reporting 0 per cent , rather than excluding them from the histogram altogether.

this histogram if our first cell could be labelled as '0' values, while the other cells are labelled with their value ranges. We will now fix both of these issues by replacing the default x-axis with a customized one.

Step 2: To customize an x-axis, we first need to produce a histogram without one, using **xaxt="n"** in the following code in *R*:

```
hist(values, breaks=c(seq(-5,60,5)),xaxt="n",
  ylab= "Frequency",xlab"Predicted extinction rate
  %",main=NULL)
```

Note that we have kept in the x-axis label (**xlab**) even though the axis itself is no longer drawn in. Also, we used the **seq** argument to set our breaks more concisely than in Step 1.

Step 3: Next we create a list of our bin ranges for the x-axis labels:

```
xlabels <- c("0","0-5","5-10","10-15","15-20","20-25","25-30",
"30-35","35-40","40-45","45-50","50-55","55-60")
```

Step 4: Now we can add our x-axis labels to the histogram already plotted using R's **axis** function:

```
axis(side=1, at=c(seq(-2.5,57.5,5)), labels=xlabels,tick=
FALSE,las=2,line=-1)
```

Breaking this code down in order, we are telling R we want an axis:

- At **side** '1' of the figure. In R, the sides of plots are numbered: 1=bottom, 2=left, 3=top, 4=right.
- With labels positioned **at** the values in the **sequence** between -2.5 and 57.5 at intervals of 5. Notice that these are not the same minimum and maximum values as we used to draw the histogram—this is because we want our labels positioned in the middle of each bin rather than at the ends (e.g. -5/2 = -2.5).

- With the **labels** we defined in our list **xlabels**.
- With no **tick** marks or axis line drawn (hence, FALSE). Here we're going to position our labels centrally underneath each cell, but you could keep ticks (TRUE) if you like.
- With the labels perpendicular to the x-axis (**las=2**) so that they don't all overlap each other. **las** indicates the orientation of tick mark labels (and other text), where: 0 (default)=always parallel to the axis, 1=always horizontal, 2=always perpendicular to axis, 3=always vertical.
- With the labels positioned close to the cells, and not overlapping the overall x-axis label—we do this with **line=-1**. This refers to the number of lines into the plot margin the axis labels will be drawn at (see section 7.2 for more details on plot margins).

For more details on the **axis** function and how to customize axes see section 7.5.2.

In Figure 4.6, we have corresponding bin labels positioned smartly at the centre of each cell, and with a shorter y-axis range we can now see the patterns of studies reporting >0 per cent extinction rates more clearly. Next, adding grid lines might make the frequencies of each cell even easier to read, though this is not needed if your aim is simply to show a general trend, as is often the case for histograms (we do so here to demonstrate the code). Adding colour to the cells will also improve the figure's visual appeal.

Step 5: If you wanted to add grid lines to your histogram, the first step is to produce a colourless version of the plot, specifying the breaks but without including either axis (**xaxt="n"**, **yaxt="n"**) or axes labels (**xlab= ""**,**ylab=""**):

```
hist(values, breaks=c(seq(-5,60,5)),xaxt="n", yaxt="n",xlab=
"",ylab="",main=NULL)
```

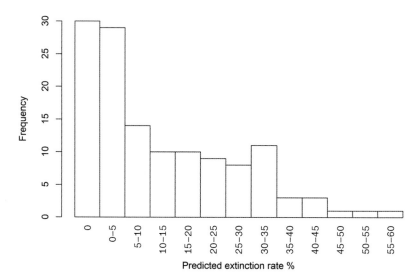

Figure 4.6 A histogram showing the distribution of averaged percentage predicted extinction rates across 130 published studies (Urban 2015), with specified bins and a customized x-axis showing bin ranges.

Step 6: We then need to add in horizontal grid lines using the function **abline** at suitable, rounded intervals given the maximum height of the y-axis we have seen earlier (30):

```
abline(h=(seq(0,30,1)), col="lightgray")
abline(h=(seq(0,30,5)), col="darkgray")
```

h=(seq tells R which axis values we want a sequence of horizontal lines at. We add minor grid lines, which we colour **"lightgray"** and major grid lines which we colour **"darkgray"**. See chapter 3 and section 7.3.4 for other examples of using **abline** to add grid lines to plots.

   Step 7: Now we need to tell R that we want to overlay this stripped-down existing plot of grid lines with a new plot, using the short line of code:

```
par(new=TRUE)
```

Step 8: And we can finally draw our high-quality histogram from before on top of the current plot of grid lines with the code below, this time adding colour too. Most of the commands and arguments here will be familiar to you from steps 2–5:

```
hist(values, col=c("mediumpurple4",rep("seagreen2",
12)),breaks=c(seq(-5,60,5)),xaxt="n",
  ylab= "Frequency",xlab="Predicted extinction rate
  %",main=NULL)
xlabels <- c("0","0-5","5-10","10-15","15-20",
  "20-25","25-30","30-35","35-40","40-45","45-50",
  "50-55","55-60")
axis(side=1, at=c(seq(-2.5,57.5,5)), labels=xlabels,tick=
  FALSE,las=2,line=-1)
```

We used the argument **col** to bring colour to Figure 4.7, and instead of choosing only one colour for all the cells we provided R with a list of colours

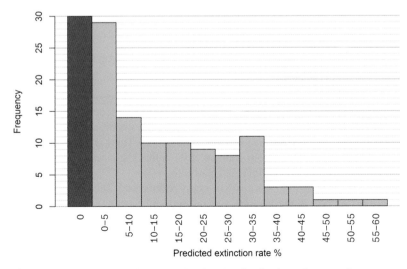

Figure 4.7 A high-quality histogram showing the distribution of averaged percentage predicted extinction rates across 130 published studies (Urban 2015), with a customized x-axis showing bin ranges, grid lines, and colour. The cell showing studies reporting 0 per cent extinction rates is coloured purple.

in order to differentiate the initial cell showing studies reporting 0 per cent extinction rates. After naming the colour for the initial cell ('**mediumpurple4**'), we used the **rep** function to add 12 of the colour name '**seagreen2**' to our **col** list for all of the remaining cells of the histogram. We now have a high-quality histogram with a sensible number of bins, that is easy to read and appealing to look at.

Before we finish with histograms, we want to mention one final refinement. All the histograms we have considered show frequencies on the y-axis. You could instead rescale so that the areas of all the rectangles add up to 1—this is called a frequency density histogram. The proponents of this approach feel that this makes comparisons within and between charts easier; we are not especially convinced, but if you would like to investigate more we provide some resources in the Further Reading at the end of this chapter.

 Key point

Histograms can be highly effective as a way to see all of the data for a single variable in a single sample, and R can help you draw really clear and attractive histograms. Often the trickiest issue with a histogram stems from your freedom to choose how to organize the data into bins, but this just needs a little care to get right.

4.4   Boxplots

4.4.1 Simple boxplot

Like histograms, boxplots can also give a good sense of the spread of data, but additionally they make some key descriptive statistics easy to interpret.

Step 1: As with histograms, as soon as you have loaded the data from section 4.2 into R it is easy to quickly visualize our list of predicted extinction rates, **values**. Try running the code:

```
boxplot(values, ylab= "Predicted extinction rate %")
```

As before, **ylab** tells R the y-axis label, but note that we have now labelled the y-axis (rather than the x-axis) with the predicted extinction rates and have not included the argument **xlab**. This is because boxplots ordinarily plot the measured variable up the y-axis rather than along the x-axis, and the box for a single sample occupies only one location at the x-axis. There is also no need for us to include **main=NULL** in our code because, unlike histograms, the **boxplot** function does not by default include a plot title.

A boxplot such as Figure 4.8 is fairly easy to interpret. The dark line in the middle of the box is the median, and the top and bottom edges of the box are the third and first quartiles respectively. We provide a recap of simple descriptive statistics in Bigger Picture 4.2. The interquartile range (IQR) can be easily calculated by subtracting the first quartile from the third quartile; IQR is a great descriptive measure for the spread of the middle 50 per cent of your values. But the whiskers protruding from the top and bottom of the box can be a little more tricky to understand, so we will discuss them in more detail next.

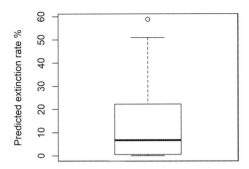

Figure 4.8 A simple Tukey boxplot showing the distribution of averaged percentage predicted extinction rates across 130 published studies (Urban 2015).

Bigger Picture 4.2
Simple descriptive statistics

Below we describe some descriptive statistics that can be used to summarize a sample of data, beyond use in graphing alone:

- Median: The 'middle' value of a data set, with an equal number of values above and below it when the values are ordered in size. This is visualized in BP 4.2 Figure A. The median will be a set value when the data set has an odd number of values, but may need to be calculated as the mean of the two middle numbers in data sets with an even number of values.

- Mean: The calculated 'characteristic' value of a data set, calculated by adding up all the values and dividing the total by how many values there were in the data set.

- First (lower) quartile: The middle value between the smallest value in the data set and the median, as indicated in BP 4.2 Figure A.

- Third (upper) quartile: The middle value between the highest value in the data set and the median, as indicated in BP 4.2 Figure A.

- Interquartile range (IQR): The range of values in the middle 50 per cent of the data set, calculated by subtracting the first quartile from the third quartile. The span of IQR is visualized in BP 4.2 Figure A.

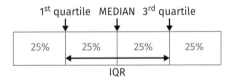

BP 4.2 Figure A A visualization of where in a list of data values some descriptive statistics lie, when the values are ordered chronologically and split into four sections (25 per cent each).

4.4.2 What do the 'whiskers' of a boxplot represent?

If you are using R to plot your boxplot, the whisker at the top of the box extends to **either**:

1 the data point furthest from the median that is still 1.5 times the interquartile range (IQR) from the top (upper quartile) edge of the box

2 the maximum value of the data set.

It will be whichever of these results in the *shortest* whisker.

Similarly, the whisker at the bottom of the box extends to **either**:

1   the data point furthest from the median that is still 1.5 times the interquartile range (IQR) from the bottom (lower quartile) edge of the box

2   the minimum value of the data set.

Again, whichever of these results in the *shortest* whisker is adopted.

These whisker rules sound somewhat overly complex, but the whiskers are basically there to indicate variability outside the upper (third) and lower (first) quartiles (i.e. among the very highest and lowest values in your sample). Different types of boxplot have different rules for plotting whiskers, but R uses Tukey boxplots and they follow the rules we give above; we discuss some of the variation in what whiskers can show in Scientific Approach 4.1. In Figure 4.8, there are no data points plotted below the bottom whisker, so we can assume this whisker reaches to the minimum value of the data set (this should also seem obvious given that the whisker reaches to 0 and we know our data does not include negative values). There is, however, a data point plotted above the top whisker (the unfilled circle), which suggests that the top whisker does not reach to the maximum value of the data set but rather the data point furthest from the median that is still 1.5 times the interquartile range from the upper quartile.

Scientific Approach 4.1
Rules for whiskers

Different scientists take different approaches to the type of whiskers that they use. We will take you through some of the alternatives. However, we prefer Tukey's definition, which is also the default in R.

The whiskers on a boxplot in R follow particular rules, such that the upper whisker is *either* the maximum value *or* largest data point less than the third quartile PLUS (1.5 x the IQR) (whichever leads to the shorter whisker); and the lower whisker is the minimum value or smallest data point that is still more than the first quartile MINUS (1.5 x the IQR) (whichever leads to the shorter whisker). These rules are assumed by default in R because R automatically uses Tukey's boxplot. However, if a different software package is used to draw a boxplot then you can get variations in what the whiskers of boxplots show. Some show:

- exclusively the minimum and maximum data values

- one standard deviation above and below the mean of the data

- the 9th and 91st percentiles

- the 2nd and 98th percentiles.

Tukey boxplots are very common and present data well. Their rules for whiskers can be very good at helping readers identify outliers because where odd data points have extreme values the shortest whiskers will most likely be the '1.5 x IQR' convention whiskers, such that outlying values will be plotted outside the confines of the box-and-whiskers. Because of the variability in conventions used in whisker plotting, however, it is therefore useful to state in figure captions which whisker convention is being used in your plot. If you simply call your boxplot a 'Tukey boxplot', then that is all you need say to explain R's default whiskers.

The major function of whiskers concerns outlier detection; whiskers are an easy way to throw up data points that seem unusual in the data set. Any points that lie outside the whiskers can be considered unusual or noteworthy values (outliers). In Figure 4.8, for example, we can see a single unfilled circle near the top of the plot lying outside the boxplot whiskers—this is an outlier from our data set of extinction rate predictions. Outliers can result from variability in the measured variable or from experimental error. Sometimes, where outliers are due to error it is important to exclude them from the data set to avoid them skewing the other values; this is one reason why outlier identification is important.

4.4.3 Refined boxplot: editing axes, adding colour, text, and reference lines

Now that we understand what our simple boxplot shows, we can add some additional code to produce a smarter-looking boxplot in R.

Step 1: We are already familiar with how to specify colour using `col` from our histogram example in section 4.3.5, so we will also edit the y-axis slightly in the code below:

```
boxplot(values, ylab= "Predicted extinction rate %",
ylim=c(-5,70), yaxs = "i", col="cadetblue1")
```

Here, `ylim` is used to define set bounds we want the y-axis to run between. In this case, it would be useful for the y-axis minimum to sit a bit below 0 so that the whiskers of the boxplot are not obscured by the axis—we have chosen -5 as this is not so far below the 0 that there will be a big gap at the bottom. Similarly, we want to have a bit of space at the top of the plot so that the outlier data point does not feel squashed in by the top of the plot, so we have chosen a round value of 70, which is not too much greater than the maximum value of the data set. We have also included the code `yaxs = "i"` which fits neat axis labels based on the specified `ylim` range (have a look at running the code with and without this argument to see the impact it has).

Also, note that when choosing colours for boxplots, it is particularly important to select colours that will not obscure the dark line displaying the median within the box, so here we have opted for a light '`cadetblue1`'.

Step 2: You may want to add text to a plot, for example to highlight an unusual or important data point, or to provide additional information such as descriptive statistics. While it is important to avoid having too much 'chart junk' on figures (see section 1.7 for more of a discussion), we will here add some text to the plot to demonstrate how simple it is to do so in R.

The text we are going to add will identify our data's outlier and a couple of descriptive statistics, so we first need to identify those specific values using R's `summary` function:

```
summary(values)
```

We can see from this that the median of our predicted extinction rates data is 6.7 per cent, the lower (first) quartile is 0.55 per cent, and the upper (third) quartile is 22.2 per cent. Also, we can tell that the outlier data value on Figure 4.8 must be the 58.8 per cent maximum.

Step 3: Now, we can very easily add these values to the figure by using the `text` function in R:

```
text(0.65,58.8, labels="outlier = 58.8")
text(0.65,6.7, labels="median = 6.7")
text(0.65,18, labels="IQR = 0.5 - 22.2")
```

The numbers following **text(** in this code are the x- and y-coordinates where the text should be positioned. The x-coordinate is here determined by the centre of our boxplot being positioned at '1' because we only have one set of data; 0.65 is a nice distance between the box and the y-axis. The y-coordinates are set at the value of the outlier and median for their respective text labels, and the interquartile range (IQR) label is positioned at a coordinate next to the box of the plot but with enough of a gap from the median label so that both pieces of text are clear. The text to be printed on the plot is then given in the quotation marks following the argument **labels**. See section 7.3.5 for more details on adding text to figures.

Altogether, Figure 4.9 displays our data smartly, identifying the outlier value and providing easy-to-interpret descriptive statistics. Although boxplots are not typically presented overlying grid lines, R gives you the flexibility to do so by using **abline** and following essentially the same steps as for other figures types (e.g. histograms in section 4.3.5, bar charts in chapter 3, and scatterplots in chapter 6). However, we feel that you are more likely to encounter situations where including a single reference line on a single-sample boxplot might be useful to you, and so provide an example of this next.

Step 4: Imagine that we were carrying out a one-sample test and wanted to compare the descriptive statistics in our existing boxplot with the value we were testing for: for the sake of this example, let us say a predicted extinction rate of 32 per cent. To include a reference line at this value, we would use **abline** as in the command below:

```
abline(h=32,col="red",lty=2,lwd=2)
```

Here we name only one value at which a horizontal line should be drawn (rather than a sequence of values as we did in section 4.3.5), and colour the line red. We have also edited the line type (**lty**), choosing a dashed line (type 2) rather than the default solid line (type 1), and line width (**lwd**), choosing a slightly

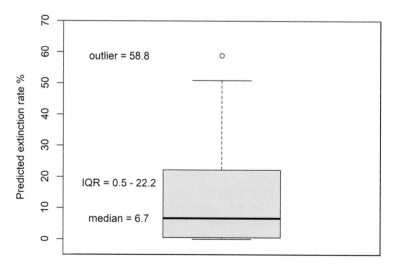

Figure 4.9 A high-quality Tukey boxplot showing the distribution of averaged percentage predicted extinction rates across 130 published studies (Urban 2015), with an edited y-axis, colour, and text labels for the outlier and key descriptive statistics.

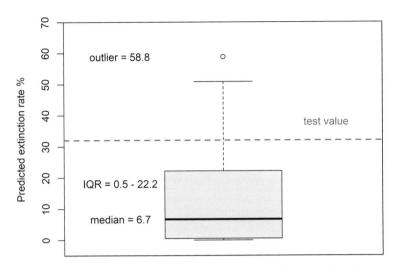

Figure 4.10 As Figure 4.9 but with added labelled reference line at predicted extinction rate 32 per cent.

thicker line (size 2) rather than the default thickness (size 1) to help it stand out; see section 7.3.4 for more details on line customization in R.

Step 5: We can also include a text label for our reference line, using the **text** function as in step 3:

```
text(1.35,38,labels="test value",col="red")
```

Again the x-coordinate value, here 1.35, was chosen as a location that positioned the text with a fairly even amount of space between the boxplot and the edge of the plotting area, based on the centre of the boxplot being situated at '1'. We also coloured the text red to match the reference line. The resulting Figure 4.10 allows clear visual comparison of our data with the imagined test value.

There are other variations on boxplots you might encounter, some of which we describe in Scientific Approach 4.2; but we think there are few circumstances where these are more useful for presenting data than the core components of boxplots we have covered.

Scientific Approach 4.2
Additional features and boxplot alternatives

There are a variety of ways people have adapted the look of boxplots, sometimes aiming to increase the information shown and sometimes to shift the focus from the centre and spread of the data to the shape of the data. Different scientists take different views on how useful these alternatives are. You may come across boxplots

with additional features or alternative figure types with components such as:

- Notches: Notched boxplots narrow the box around the median, usually to show the confidence interval around the median. This may provide indications of the statistical significance of any differences between the medians of multiple boxplots, but does not add much value to an individual boxplot displaying single-sample data.

- Overlaid strip-charts (aka strip plots, dot plots): This is where a boxplot is overlaid with all the individual data points in a strip. Showing all of the data values as well as the data summary (the boxplot) can make a figure look messy, particularly if there are more than 20 data points. See section 6.5 for more details on, and an example of, strip-charts.

- Violin plots: These are similar to boxplots except they also show the smoothed probability density of the data at different values. This means they provide more information about how the values in the data are distributed, but also that they are much noisier than boxplots.

Generally, we think that adding overlaid strip-charts or plotting your data as a violin plot is only really helpful where your data has more than one mode (i.e. the distribution is bimodal or multimodal) because, by providing a summary, a boxplot alone would obscure this. If this is the case for your data and you have a lot of data points, a strip plot will often look messy and you would likely be better off with a violin plot. If you have fewer than 20 data points, then a strip plot alone might be best because the fitted lines of a violin plot would not have many data points to be fitted to.

 Key point

Boxplots may not be as immediately intuitive as histograms, but they avoid the arbitrary binning issue, and (as we will see in chapter 5) can be really effective when you want to display and compare several samples (**multiple-samples data**).

4.5   So which should I choose? Histograms vs boxplots

We have demonstrated two main graphing types we can use to present data from a single sample: histograms and boxplots. Choosing the best graph for your purpose depends on what you want to present from your data:

- Histograms give a great sense of the distribution of a variable, are easy to interpret, and are visually appealing, but can take a little bit of tinkering to get the binning of the data right. If you are carrying out any statistical analysis of your data, it is very often valuable to visualize the data distribution of the variable of interest in advance using a histogram.

- Boxplots are easier to produce than histograms, but might be a little less attractive and a little bit harder for the lay reader to interpret. They do, however, provide an effective summary of a distribution, including descriptive statistics, and allow for easy identification of outliers. Boxplots also really come into their own when we want to compare a number of samples, and that is what we will tackle in chapter 5.

 Key point

Which you choose will be influenced by which you expect your readership to be most comfortable with, and also the aspects of the data that you want to draw attention to.

4.6   Stretch your understanding

To consolidate the ideas we have introduced in this chapter, have a look at how the following data sets (provided in the online resources) look presented both as histograms and boxplots. Which graph type do you think best presents the data?

The first data set, called 'rat_lick_data.csv', has been adapted from a study that used rodent brief-access taste aversion (BATA) experiments to assess the bitterness of quinine, an aversive-tasting drug (Sheng et al. 2016). Rats were placed in a 'lickometer' that recorded how many licks they took from sipper tubes filled with different concentrations of quinine hydrochloride dihydrate during 8-second-long trials. The data you are provided with details only the number of licks taken during the 720 trials where the concentration supplied by the sipper tubes was 0.03mM. We are interested in the distribution of licks taken during the trials (hint: before you decide on your final figure type, have a check to see if the data looks to be unimodal or if it has more than one mode).

The second data set, called 'firefly_spermatophore_mass.csv', lists the mass (in milligrams) of the spermatophores dissected from 35 *Photinus ignitus* fire-flies. These values were inferred by Whitlock and Schluter (2014) from a figure in a paper exploring female preference for male courtship flashes (Cratsley and Lewis 2003) among *P. ignitus*.

Once you've had a go, check out the online resources to find our individual attempts at these exercises, including the code we used ('R script for Authors' attempts chapter 4'), the finished figures, our comments on our design choices, and what we think of the other author's attempt ('Authors' attempts for chapter 4').

Chapter Summary

- With single-sample data, for each individual within the sample we have one single measurement. This measurement should be continuous, or be discrete but have quite a few different values.
- Histograms and boxplots are both suitable for presenting single-sample data, but which choice is most suitable depends in part on what you intend to communicate from the data.
- Using R, the basic code for histograms is: `hist(x)`.
- Histograms are excellent for showing the distribution of a set of values, but require a bit of thinking about how best to 'bin' the data.
- Using R, the basic code for boxplots is: `boxplot(x)`.
- Boxplots can also give a good sense of the spread of data, while also providing some descriptive statistics and allowing easy identification of outliers. However, they may be a bit less intuitive for viewers to interpret, particularly the 'whiskers'.

Online Resources

The following online resources are available for this chapter at www.oup.com/he/humphreys-obp1e:

- R script for chapter 4
- extinction_data.csv

- bin_play.csv
- rat_lick_data.csv
- firefly_spermatophore_mass.csv
- Authors' attempts for chapter 4
- R script for Authors' attempts chapter 4

Further Reading

- 'Histograms' explains more about some of the conceptual and presentational considerations when constructing a histogram:

  https://statistics.laerd.com/statistical-guides/understanding-histograms.php

- 'Frequency Density Histograms': in frequency density histograms, cells are rescaled so that the total area of the cells adds up to 1. This can allow for easy comparison of the relative proportions each cell contributes, even when the bin ranges and cell widths may be uneven. You can find more information on this in a Bitesize guide on Histograms on the BBC website, a glossary entry on the Underground Mathematics website, and an explanation of the difficulty of interpreting the vertical scale in histograms. Here are the links to these websites:

  https://www.bbc.co.uk/bitesize/guides/zspfcwx/revision/1

  https://undergroundmathematics.org/glossary/frequency-density

  https://modelassist.epixanalytics.com/display/EA/
  Difficulty+of+interpreting+the+vertical+scale

- 'Understanding Boxplots' explains the anatomy of a boxplot in some detail:

  https://towardsdatascience.com/understanding-boxplots-5e2df7bcbd51

- 'Boxplots' uses examples to demonstrate how to use two different kinds of boxplot:

  https://bolt.mph.ufl.edu/6050-6052/unit-1/
  one-quantitative-variable-introduction/boxplot/

- 'Notched Box Plots' shows how this variant of a boxplot can give useful extra information:

  https://sites.google.com/site/davidsstatistics/davids-statistics/
  notched-box-plots

- 'Graphs in R—Overlaying Data Summaries in Dotplots': this blog sets out how to use R to overlay a boxplot on a dot plot or strip-chart:

  https://www.r-bloggers.com/
  graphs-in-r-overlaying-data-summaries-in-dotplots/

- For more on when and how to use violin plots, try 'A Complete Guide to Violin Plots':

  https://chartio.com/learn/charts/violin-plot-complete-guide/

- 'Violin Plots 101: Visualizing Distribution and Probability Density' offers clear explanations and examples of different ways to use these diagrams:

  https://mode.com/blog/violin-plot-examples/

- 'Stem-and-Leaf Plots': you might find these plots mentioned as an additional way that single-sample data can be presented, but we would advise against their usage. Stem-and-leaf became popular as a way to make figures using text alone back when computers had no graphics facility, but it is now easy to present data in much more intuitive and visually appealing ways. If you are still interested, you can find further information on stem-and-leaf plots on the Purplemath website:

  https://www.purplemath.com/modules/stemleaf.htm

5 COMPARING MULTIPLE SAMPLES: BOXPLOTS AND HISTOGRAMS

Learning objectives

By the end of this chapter you should be able to:

- Explain the advantages and disadvantages of boxplots and histograms when comparing multiple samples.
- Produce clear and effective boxplots and histograms from multiple-samples data.
- Present grouped multiple-samples data effectively.

5.1 Introduction: when do I have multiple-samples data and what can I do with it?

In the previous chapters, we have seen effective ways to present data from a single sample. Now we turn to looking at cases where we have two or more related samples, with each sample from a different group of individuals. Like chapter 4, for each individual we are still just interested in a single measurement. Again like chapter 4, this measurement should be continuous, or have quite a lot of different possible values. We want to see if there appears to be a difference in this measurement between individuals in the different groups from which the samples are drawn. If you have discrete data with seven or fewer different values, then consider the earlier chapters (chapters 2 and 3) for dealing with qualitative data (aka categorical data) instead. We provide an example of such decision-making in the online resource 'Authors' attempts for chapter 5' when looking at the file 'nova_scotia_birds.csv' described in section 5.7.

There are three ways we can graph data from multiple samples in order to compare them: boxplots, histograms, and bar charts. As we discuss in section 5.5 (see also links in the Further Reading recommended at the end of this chapter), we think bar charts will rarely be the best option for quantitative data; so we concentrate on boxplots and histograms in this chapter. That said, you might well disagree with us on bar charts (their continuing popularity suggests that many scientists do), so we cover them in a little more depth in 'Bar charts for quantitative multiple-samples data' in the online resources for this chapter.

This chapter builds on the earlier chapters where single-sample examples of boxplots and histograms have been covered (see chapter 4), and is written assuming you have good familiarity with that material.

> **Key point**
>
> You will find histograms, bar charts, and boxplots used to show multiple-samples data. We will cover all three, but try and convince you that boxplots will almost always be the best option.

5.2   An example of multiple-samples data

The downloadable .csv file 'pig_litters' gives the birth weights (lbs) of Poland China pigs from seven different litters (adapted from Snedecor 1956). Poland China pigs are the oldest breed of pig in the USA and are known for their large size. In fact, the largest pig known, 'Big Bill', was a Poland China, weighing in at 1158 kg (2552 lb) (Wilsdon 2009). We are interested in comparing birth weights across the different litters, and considering whether birth weights differ between large and small litters.

Once you have loaded in (see the 'R Basics' guide from the online resources for chapter 1 for a reminder of how to do this) and viewed this data (we have named ours **piglets**), you will see that this is a slightly more complicated file than those we have looked at in previous chapters. There are multiple samples, each shown in their own column, and there are some NA values, which indicates that the samples vary in sample size (i.e. unsurprisingly, not all litters have the same number of piglets). This could be an issue here as R expects that we will have equal numbers of individuals in each group (so all the columns in our Excel spreadsheet file will be the same length), but actually we do not. What R does is fill out the shorter groups with null values (NA). Try running this code, choosing a new name for the final data set (**cleanpiglets**):

```
cleanpiglets <- lapply(piglets, function(col)col[!is.na(col)])
```

This code just strips out the null values in our data. Use this line any time you load a data file with columns of unequal length. We know that to many of you, this line will look very weird indeed. If it does, then your choice is just to accept that it works and use it (like we do with lots of devices), or find someone to unpack it for you. As we said right at the start of the book, we want to keep this as a graphing book and not let it grow into an R manual. Either way, if you now have a look at your data:

```
View(cleanpiglets)
```

you will see that R has now converted the data into list form and understands the sample size of each litter. We are now ready to produce some figures.

Note: Multiple-samples data will not always be formatted as lists of values in separate columns for each group. See later examples in the chapter and the additional data sets at the end for ways of extracting out the information you want to present from data in different formats.

Key point

You can save yourself difficulties down the line with a little bit of data cleaning before you start to work with the data.

5.3   Boxplots

5.3.1   Simple boxplot

In chapter 4 we produced a boxplot from single-sample data, and we can easily produce a boxplot of multiple-samples data in much the same way once our data set is in an appropriate form. Because of the way our data is arranged, in the code below we simply tell R the name of the entire data set 'cleanpiglets' and it will assume that each of the columns represents a different sample of birth weights to be plotted:

```
boxplot(cleanpiglets,ylab = "Birth weight (lb)", ylim=c(0,5),
yaxs = "i", col="plum")
```

Note: Other arrangements of multiple-samples data might require a `boxplot(y~x)` format for the command. For examples of this see section 5.6.2 and the 'Authors' attempts for chapter 5' and associated 'R script for Authors' attempts chapter 5' in the online resources.

While we specify y-axis limits (`ylim` and `yaxs="i"`) and label (`ylab`), we did not specify any x-axis arguments even though we have multiple boxes to plot. You will see in Figure 5.1, however, that R has automatically filled in our x-axis tick labels with the various column names. Note that you could also include a title using the argument `main` (though most scientific figures are accompanied

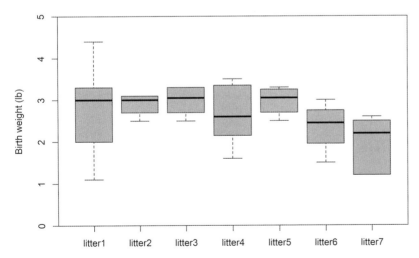

Figure 5.1 Birth weights (lbs) of Poland China pigs from seven different litters (adapted from Snedecor 1956).

by captions instead; see section 1.2 and Scientific Approach 1.2) or list multiple colours using **col**, but we would advise against using different colours unless they provide additional information not otherwise shown. We will see cases where this is more appropriate later on in the chapter when we look at overlaying plots (section 5.4.2) and grouped multiple-samples data (section 5.6).

5.3.2 Refined boxplot: adding sample sizes as text, switching order, and tidying x-axis

While our simple boxplot in Figure 5.1 does give us an indication of the differences in birth weights between litters, we cannot actually see the number of piglets in each litter. To help us discern whether birth weights differ between large and small litters, we will need to present the sample sizes of each litter alongside their corresponding boxes.

Step 1: First off, we'll need to create a list containing the sample size of each litter. You could just type these into a list yourself if you already knew them (in fact, our cleaned-up data set shows the sample size of each litter in square brackets). But we're going to follow some steps that get R to do the counting for us, to demonstrate the use of the **length** function. We'll start by creating shorter names for each litter, for ease:

```
lit1<-cleanpiglets$litter1
lit2<-cleanpiglets$litter2
lit3<-cleanpiglets$litter3
lit4<-cleanpiglets$litter4
lit5<-cleanpiglets$litter5
lit6<-cleanpiglets$litter6
lit7<-cleanpiglets$litter7
```

Step 2: Now we use the function **length**, which gives us the number of values in the data named. If we use it for each litter, we can list the sample sizes consecutively under the list name 'n' with:

```
n <- c(length(lit1),length(lit2),length(lit3),length(lit4),
length(lit5),length(lit6),length(lit7))
```

Have a look at **n** by running it through the Console, and we can see it lists our sample sizes in order for the litters 1–7 as: 10, 4, 8, 8, 8, 4, 6.

Step 3: It is now really easy to add these sample sizes to the existing boxplot (like the one in Figure 5.1) using the text function:

```
text(x= c(1,2,3,4,5,6,7), y= 0.5, labels= c(n), col=
"purple4")
```

Here, in **x** we list the x-coordinates (simply 1–7 as we have seven samples plotted), in **y** we say at what height up the y-axis we want the text (this could also be a list if you wanted them at different heights), and **labels** is where we say what we want the text to say—here we want the values listed in 'n' to be positioned at the respective x-coordinates.

Overall, in Figure 5.2 it looks like there is not much notable difference between the birth weight medians across litters, but there is quite a bit of within-litter variation (particularly in litter 1). Litter size does not seem to have any clear correlation with larger or smaller birth weights. However, this might be clearer still if we ordered the boxes by litter size.

Step 4: To visually emphasize the issue of sample size in the figure, we will order the boxes by increasing litter size. To do this, we first create a new list (`switcheddata`) where we tell R we want the samples from our `cleanpiglets` data ordered starting with the list that currently constitutes the second column (litter 2), adding the other lists in order of increasing sample size (litters—or columns—6, 7, 3, 4, and 5), and ending with the list that currently constitutes the first column (litter 1) and has the largest sample size:

```
switcheddata <- cleanpiglets[c(2,6,7,3,4,5,1)]
```

Note the use of square brackets when specifying the positions of samples within the `cleanpiglets` data. If you now view `switcheddata`, you will see how the order has changed as specified compared to `cleanpiglets`.

Step 5: We can also easily switch round the order of our list of sample sizes, by using the same trick on our existing list `n`:

```
switchedn<- n[c(2,6,7,3,4,5,1)]
```

Step 6: We can now simply substitute `switcheddata` into the boxplot code from earlier:

```
boxplot(switcheddata,ylab = "Birth weight (lb)", ylim=c(0,5),
yaxs = "i", col="plum")
```

Step 7: And add our `switchedn` sample sizes as text below the corresponding boxes:

```
text(x= c(1,2,3,4,5,6,7), y= 0.5, labels= c(switchedn),
col= "purple4")
```

Note: If you wanted to present sample sizes on a figure preceded by 'n=', see sections 2.2.1.2, 2.2.2, 6.3.3, and 8.6, the online resource 'Bar charts for quantitative multiple-samples data' for this chapter, and Scientific Approach 8.1 for how the `paste` function can be used to present values alongside text.

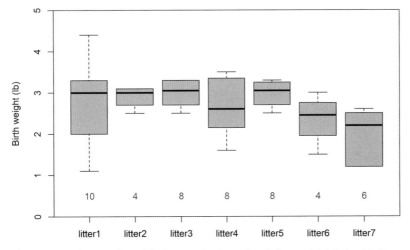

Figure 5.2 As Figure 5.1 but with the sample sizes of each litter added below their respective boxes.

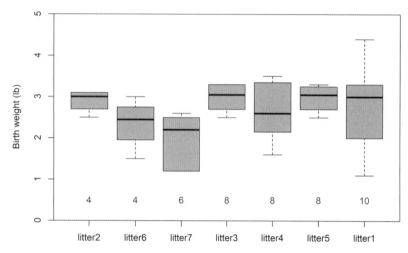

Figure 5.3 As Figure 5.2 but with the boxes and respective sample sizes ordered by increasing litter size.

Step 8: Although R is still correctly filling in our x-axis tick labels (given that we switched the order) in Figure 5.3, it takes the formatting of the identifiers of different litters from the column titles of the original data file, which does not look as tidy as it could. To neaten these x-axis tick labels, we can rename them manually using the **names** argument as in the boxplot code below:

```
boxplot(switcheddata,ylab = "Birth weight (lb)", ylim=c(0,5),
  yaxs = "i", names=c("Litter 2", "Litter 6", "Litter 7",
  "Litter 3", "Litter 4", "Litter 5", "Litter 1"), col="plum")
text(x= c(1,2,3,4,5,6,7), y= 0.5, labels= c(switchedn),
  col= "purple4")
```

Alternatively, we could have used **names** to make the x-axis tick labels as numbers only and added an x-axis label (using **xlab**) called 'Litter'. Either way, we are left with a high-quality boxplot, as shown in Figure 5.4, which is intuitive to interpret and makes comparisons between the birth weights of different litters easy. There really doesn't seem to be an obvious effect on the birth weight of a pig due to the size of the litter it was born into.

 Key point

Boxplots extend very naturally to accommodate multiple-samples data, and all the tweaks that we explored for single samples will still be available.

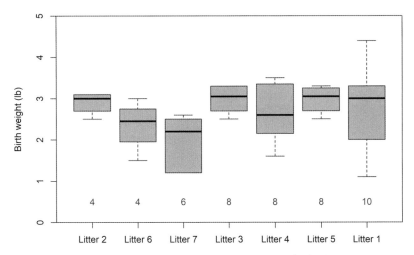

Figure 5.4 A high-quality boxplot showing the birth weights (lbs) of Poland China pigs from seven different litters (adapted from Snedecor 1956), in order of increasing litter size (shown in dark purple text below boxes).

5.4   Histograms

If we instead wanted to present the data from multiple samples in histogram form, we essentially have two options: i) plot multiple histograms, one for each sample, alongside one another, or ii) sit multiple histograms on top of each other in a single plot. As you might imagine, the first option could take up a lot of space on a page if you have many samples and it would be difficult to make comparisons between plots, while the second option could look very messy with lots of samples layered on top of each other. For this reason, we will demonstrate how to plot only three of the sample litters from our Poland China pigs for both options. Here we'll look at the three litters with the greatest interquartile ranges, or IQR (the biggest boxes in our boxplot): litters 1, 4 and 7. (See Bigger Picture 4.2 in chapter 4 for an in-depth explanation of IQR.)

5.4.1   Multiple histogram plots

Step 1: To create multiple plots at the same time in the plot window of R, we have to use **par(mfrow=)** as we did when playing with bins in chapter 4 (see also section 7.2 for more uses of **par**). Remember, the numbers listed in this function refer to the number of rows and columns we want the plots arranged in. So, first let's arrange our three histograms vertically (that is, three rows in one column), using the short litter names we created in section 5.3.2:

```
par(mfrow=c(3,1))
hist(lit1, col="darkgoldenrod1",xlim=c(1,5), xaxs = "i",
  ylim=c(0,4), yaxs = "i",
  xlab="Birth weight (lb)",main="Litter 1")
hist(lit4, col="darkgoldenrod1",xlim=c(1,5), xaxs = "i",
  ylim=c(0,4), yaxs = "i",
  xlab="Birth weight (lb)",main="Litter 4")
hist(lit7, col="darkgoldenrod1",xlim=c(1,5), xaxs = "i",
  ylim=c(0,4), yaxs = "i",
  xlab="Birth weight (lb)",main="Litter 7")
```

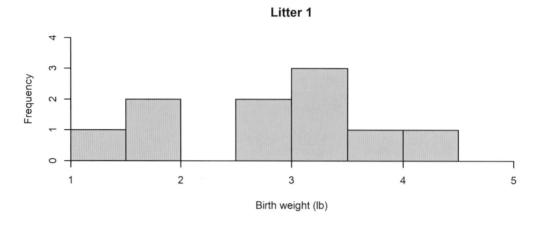

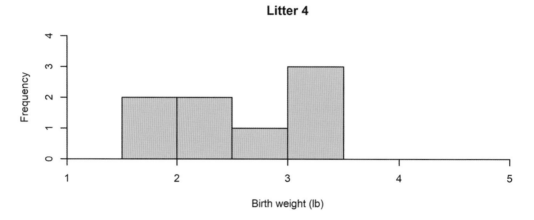

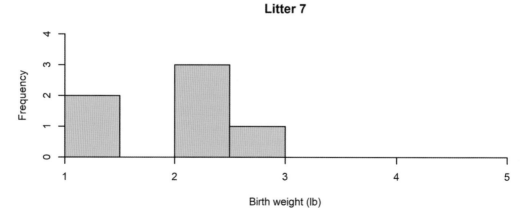

Figure 5.5 Distributions of birth weights (lbs) of Poland China pigs from litters 1, 4, and 7 (adapted from Snedecor 1956). The vertical arrangement makes comparison of birth weights easy, but comparison of frequencies difficult.

Note that in the code for Figure 5.5 we made the y-axis and x-axis span the same ranges on each figure in order to aid comparison. We also gave each plot a title that identified its corresponding litter. However, while the aligned x-axes in Figure 5.5 make comparison of the birth weights easier, the vertical arrangement is not very helpful for comparing frequencies.

Step 2: Alternatively, we could present our three histograms arranged horizontally (in one row, but with three columns), as in the code below:

```
par(mfrow=c(1,3))
hist(lit1, col="darkgoldenrod1",xlim=c(1,5), xaxs = "i",
  ylim=c(0,4), yaxs = "i",
  xlab="Birth weight (lb)",main="Litter 1")
hist(lit4, col="darkgoldenrod1",xlim=c(1,5), xaxs = "i",
  ylim=c(0,4), yaxs = "i",
  xlab="Birth weight (lb)",main="Litter 4")
hist(lit7, col="darkgoldenrod1",xlim=c(1,5), xaxs = "i",
  ylim=c(0,4), yaxs = "i",
  xlab="Birth weight (lb)",main="Litter 7")
```

Conversely to Figure 5.5, the horizontal arrangement of Figure 5.6 makes comparison of frequencies easier, but comparison of birth weights between litters slightly more difficult. Arranging histograms vertically or horizontally appears to make *either* birth weights *or* frequencies easily comparable, but not both. Also, both Figure 5.5 and Figure 5.6 take up more space on the page than a single plot would, because they need to be stretched so that the axes of all three plots are readable.

5.4.2 Overlaying histograms in a single plot

Step 1: We have seen above that plotting multiple figures can take up a lot of space, so we will now look at overlaying histograms in a single plot. To do this, we first need to tell R that we want to return to looking at one figure in one column and one row:

```
par(mfrow=c(1,1))
```

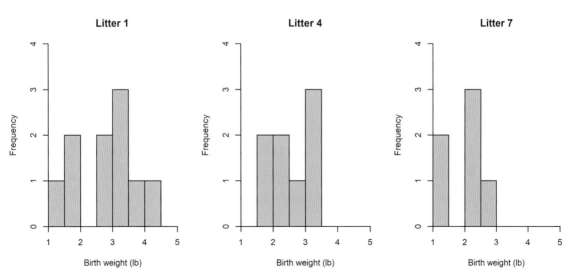

Figure 5.6 As Figure 5.5 but with plots arranged horizontally. The horizontal arrangement makes comparison of frequencies easy, but comparison of birth weights difficult.

Step 2: Now we can plot our three histograms, but it is important (particularly now we are overlaying them) that we set their bins to span the same widths using the **breaks** argument introduced in chapter 4:

```
hist(lit1, breaks=c(seq(1,5,0.5)), col="yellow2",xlim=c(1,5),
  xaxs = "i", ylim=c(0,4), yaxs = "i",
  xlab="Birth weight (lb)",main=NULL)
par(new=TRUE)
hist(lit4, breaks=c(seq(1,5,0.5)),col="springgreen3",
  xlim=c(1,5), xaxs = "i", ylim=c(0,4), yaxs = "i",
  xlab="",ylab="",main=NULL)
par(new=TRUE)
hist(lit7, breaks=c(seq(1,5,0.5)),col="royalblue3",
  xlim=c(1,5), xaxs = "i", ylim=c(0,4), yaxs = "i",
  xlab="",ylab="",main=NULL)
```

A key piece of code above when plotting the second and third histograms is **par(new=TRUE)**—this was used to add plots on top of grid lines in chapters 3 and 4. Once the first histogram is plotted, we tell R that we want to just add the further histograms on top, so that the plots overlay each other. Note that in the second and third histograms we specify blank x- and y-axis labels (**lab**) and no titles (**main**), because the base plot already included the axis labels and we do not need a title on this plot (we instead provide a detailed caption—see section 1.2 and Scientific Approach 1.2 for advice on writing captions). Also key here is the use of different colours in each histogram; as the plots will be placed on top of each other, colour is necessary to distinguish the data from the different litters. However, this does also mean we'll need to add a legend to explain the different colours.

Step 3: As covered in chapters 2 and 3, adding a legend to a figure is simple:

```
legend("topright",title="Litter", legend=c("1","4","7"),
fill=c("yellow2","springgreen3","royalblue3"), bty="n")
```

As a reminder, here **topright** tells R where to position our legend, **title** is what we want the heading of the legend box to be, in the argument **legend** we list the samples plotted, **fill=c(** lists the corresponding colours, and **bty="n"** tells R we do not want the legend to be encased in a box (see section 7.4 for more legend customization tips).

Although the resulting histogram (which you can see in Figure 5.7) looks neatly presented, some values are obscured for some samples; only litter 7 (the last one we added to the plot) is entirely visible. This makes it a far less effective plot than the three-panel Figures 5.5 and 5.6 as it does not communicate all three samples fully. One way we could try to address this problem and allow for better comparison between samples is by making the histogram cells slightly transparent.

Step 4: To customize the transparency of colours, we first need to install and activate the package 'yarrr' (Phillips 2017). You can do this via your preferred method (see the 'R Basics' guide from the online resources for chapter 1 for a recap), but we have provided the code to do this below:

```
install.packages("yarrr")
library(yarrr)
```

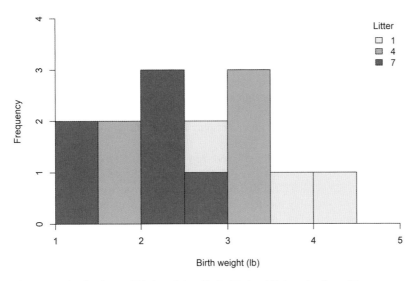

Figure 5.7 Distributions of birth weights (lbs) of Poland China pigs from litters 1 (yellow), 4 (green), and 7 (blue) (adapted from Snedecor 1956). Not all values for litters 1 or 4 are visible.

Step 5: The 'yarrr' package gives us access to the function **transparent**, which makes altering the transparency of colours easy. We simply have to enter the original colour as the first argument **orig.col**, then enter how transparent we want to make it as the second argument **trans.val**, from 0 (opaque) to 1 (fully transparent). Try out the code below, running each histogram one at a time to see how they stack on top of the other:

```
hist(lit1, breaks=c(seq(1,5,0.5)), col=transparent(orig.col
  = "yellow2", trans.val = 0.4),xlim=c(1,5), xaxs = "i",
  ylim=c(0,4), yaxs = "i",
  xlab="Birth weight (lb)",main=NULL)
par(new=TRUE)
hist(lit4, breaks=c(seq(1,5,0.5)),col=transparent(orig.col
  = "springgreen3", trans.val=0.6),xlim=c(1,5), xaxs = "i",
  ylim=c(0,4), yaxs = "i",
  xlab="",ylab="",main=NULL)
par(new=TRUE)
hist(lit7, breaks=c(seq(1,5,0.5)),col=transparent(orig.col
  = "royalblue3", trans.val=0.8),xlim=c(1,5), xaxs = "i",
  ylim=c(0,4), yaxs = "i",
  xlab="",ylab="",main=NULL)
```

Step 6: Again, we can easily add a legend, though this time we also need to include our different transparency levels for each colour:

```
legend("topright",title="Litter", legend=c("1","4","7"),
fill=c(transparent(orig.col = "yellow2", trans.val = 0.4),
transparent(orig.col = "springgreen3", trans.val=0.6),
transparent(orig.col = "royalblue3", trans.val=0.8)),
bty="n")
```

Now in Figure 5.8, by looking closely at the colours of the cells, we can determine more information about the different litters than we could in Figure 5.7. For example, litter 1 had no piglets of birth weight between 2 and 2.5 lbs, but litter 4 had two piglets in this weight range and litter 7 had three piglets in this weight range. However, the overlapping colours are hard work to decipher and the figure is still far from being easy to interpret. We still believe that Figures 5.5 and 5.6 are more effective at presenting the data from our three sample litters, but have demonstrated how to display multiple histograms in a single plot in case it is useful to you.

Histograms can be a really effective way of presenting data to give a good sense of its distribution in a way that boxplots can't. However, in layering histograms on top of each other like this, things get messy pretty fast. Three different samples is probably the limit for effective data presentation, and even then it may depend on what the spread of values looks like—you may still be better presenting each sample separately, as in Figures 5.5 and 5.6. Our overlapping histogram could have worked a lot better if there were larger sample sizes, or if the distributions differed more notably. By contrast, boxplots do not really have this problem when presenting multiple samples—it is really easy to read data from each sample where you have several boxes-and-whiskers lined up.

> ### 💡 Key point
>
> We think boxplots are better than histograms if you want to explore several samples together. But if you really feel you are happier with histograms, there are some techniques you can apply to maximize their effectiveness with multiple-samples data. But the more samples you have, the more we think life will be easier for you and your readers if you use a boxplot.

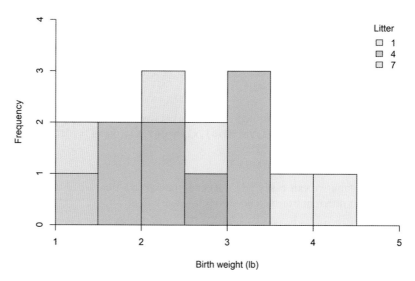

Figure 5.8 A overlaid histogram showing the distributions of birth weights (lbs) of Poland China pigs from litters 1 (yellow), 4 (turquoise), and 7 (pale blue) (adapted from Snedecor 1956).

5.5   Boxplot, histogram, or bar chart: which is best?

Boxplots (or violin plots—see Scientific Approach 4.2 and the Further Reading in chapter 4 for details and helpful links) are very good for comparing samples, as they are easy to read side by side and provide a lot of detail about each sample in a compact form. We think on most occasions they will be your figure type of choice for comparing samples of quantitative data.

As we explored in chapter 4, histograms can sometimes give a better sense of the distribution of each sample than boxplots. However, when we want to compare multiple samples it can be difficult to compare multiple histograms, and overlaid histograms will only be easy to interpret with a few samples that vary substantially in their distributions. Histograms of samples on separate axes very rarely feel like the best option to us. Multiple histograms overlaid on the same axes can only work effectively if the number of samples is low (we think two or three), and if the distributions are very different so there is little obscuring one histogram with an overlaid one.

Bar charts are generally less informative than boxplots, giving the reader less detail about each sample (see also the links in the Further Reading for how bar charts can misrepresent quantitative data). That is why we don't cover them in this chapter. However, bar charts remain popular. Also, with bar charts you get great flexibility in what you present, getting to choose whichever measures of central tendency and variability you want. But remember you can also change what the whiskers in your boxplot signify if you want, and you could add text to present the means in your boxplot. So we really don't see many (if any) situations where the bar chart will end up being your best option for comparing multiple samples of quantitative data; but if you want to explore them more in this context see the online resource 'Bar charts for quantitative multiple-samples data' and its associated R script file.

With certain data sets, you may have multiple samples that can be logically grouped together. We will briefly cover how you might want to present grouped multiple-samples data next.

 Key point

We are not at all enthusiastic about using bar charts for quantitative data. But you will often find publications where the authors use them in just that context, so clearly not everyone agrees with us.

5.6   Grouped multiple-samples data

5.6.1 An example of grouped multiple-samples data

So far in this chapter we have looked at multiple-samples data, where each sample comes from a different group of individuals, and the groups are not connected to or paired with each other in any particular way. However, sometimes you might collect data where some of the groups have more in common with each other than others, and you therefore might want to present them closer to each other or in a way that makes their link apparent. Most commonly, this is due to the inclusion of an additional variable that was measured during data collection, such that for each group there are essentially multiple sub-groups of measurements that share the group in common. For example, a study might measure the levels of serotonin in human volunteers belonging to treatment groups that are prescribed different antidepressants. Within these groups, the researchers want to compare the effects of the different treatments on males compared to females, and so when presenting their findings they will show the male and female data for each treatment group alongside one another to make it clear that the data was recorded from the same treatment.

Another example of grouped multiple-samples data comes from an experiment into how different amounts of water and covering treatments affect the germination of seeds in boxes (Chatfield 1982). In the data set 'seed_data.csv' we provide in the online resources (you might recognize this from the 'Walk-through example' from chapter 1), you will see that the data is organized in a different way to the data on pig litters from earlier in this chapter. There are separated rows for each seed box and three columns: the first describes the treatment (that is whether boxes were uncovered or covered in order to slow evaporation), the second describes the level of watering that boxes received (coded as 1–5, with 1 corresponding to the least water and 5 corresponding to the most water), and the third column states the number of seeds that had germinated in each box after two weeks. For each watering level, four uncovered and four covered boxes were sown with 100 seeds (Chatfield 1982). You might have spotted that there are only results for three boxes in the covered treatment experiencing water level 5, but this is because no result was recorded for the fourth box in the original data.

Because of the way this data is organized there are no NA values (as we dealt with in section 5.2), so this missing value is not a problem for our figure production, but it is still something we might want to mention in our figure caption (see section 1.2 and Scientific Approach 1.2 for a reminder of what makes a good figure caption). As a side-note, there is no right or wrong way to organize your data. However, the way we have presented it here with each row representing a data point is probably more common. Since all the data is stacked on top of each other, this is called stacked data organization, whereas the separate columns for each sample format used for our Poland China pigs is called unstacked. You can organize any data set in stacked or unstacked format, and R can cope either way. We would recommend you follow a stacked data format when possible, as this is usually a more logical way to structure data, but ultimately you should pick the way that seems most intuitive to you for your data.

There are two data presentation options to choose from when dealing with grouped multiple-samples data such as the results of the seed germination experiment: boxplot or bar chart. We will demonstrate our preferred boxplots here and relegate bar charts to the online resource 'Bar charts for quantitative multiple-samples data' and its associated R script file, using the name '**seeds**' for our loaded-in data set.

5.6.2 Grouped multiple-samples boxplots

To present our grouped data '**seeds**' as a boxplot we will first plot the results from one of the treatments, 'uncovered', and then add the data from the second treatment, 'covered', having made sure that there is suitable space in the original plot for corresponding boxes to be positioned adjacent to one another.

Step 1: In order to plot the data for each treatment one at a time, we first need to subset our main data set into two new data sets, each containing either all the data for the 'uncovered' or all the data for the 'covered' boxes in the experiment. To do this, we use the **subset** function, as in the code below:

```
uncovered <- subset(seeds, treatment=="uncovered")
covered <- subset(seeds, treatment=="covered")
```

'Subsetting' can be very useful wherever you just want to use a particular part of your data set for plotting or carrying out statistical tests. We tell R we want to **subset** data from the data set '**seeds**' and then say that if the '**treatment**' variable of that data equals a certain treatment ('**uncovered**' or '**covered**') we want to add it to a new data group (respectively, **uncovered** or **covered**) named in front of the arrow. It is important to note here that we are required to use the relational operator **==** (which, in R, means 'exactly equal to') rather than simply **=** (which, in R, assigns values and can be used as an alternative to the **<-** symbol).

Step 2: Now, we first plot the data for the uncovered boxes, specifying that we want to plot the number of seeds that germinated (y-axis) against (~ in the boxplot formula) the level of watering (x-axis), identifying the variables as part of the **uncovered** subsetted data using the **$**:

```
boxplot(uncovered$germinated~uncovered$water,
ylab = "Seeds germinating per box", xlab = "Amount of water",
ylim=c(0,100), yaxs = "i", xlim=c(0.7,5.3), xaxs= "i",
xaxt="n",col="gray",
boxwex=0.27,at = 1:5 - 0.15)
```

There are only a couple of arguments in the boxplot code new to this chapter. **boxwex** is a scaling factor that will be applied to all boxes in the plot; here we have chosen the value 0.27 to make the boxes narrower than they are by default, so that the boxes for the covered data will be able to fit alongside neatly. **at** tells R where we want the boxes placed along the x-axis. By default, R would position our five boxes at points 1–5 along the x-axis, but here we want each of the '**uncovered**' data boxes to be positioned off the centre of these points so that there is room for the '**covered**' data boxes around each point too. To achieve this, we have used the code **1:5 - 0.15**, which will place each box 0.15 to the left of its default 1–5 x-axis position. Because our boxes are not going to be placed at their default positions along the x-axis, but this is still where we will want our watering level labels to be positioned, we do not include any details for the x-axis yet (**xaxt="n"**).

Step 3: We can now add our covered boxes data to the plot, by preceding the main code with **par(new=TRUE)** (as seen earlier in section 5.4.2 and in chapters 3 and 4, this adds subsequent code onto existing plots), and by using a similar y~x arrangement of variables in the **boxplot** function:

```
par(new=TRUE)
boxplot(covered$germinated~covered$water,
  ylab = "", xlab = "",
  ylim=c(0,100), yaxs = "i", xlim=c(0.7,5.3),xaxs= "i",
  yaxt="n",xaxt="n",col="yellow",
  boxwex=0.27,at = 1:5 + 0.15)
```

Notice that we set the y-axis and x-axis ranges to exactly the same as the initial boxplot, and do not need to add any y- or x-axis ticks. We set **boxwex** to 0.27 again, so that the boxes are drawn to the same scale as the previous treatment, but this time when positioning our boxes we set them to 0.15 to the right of their default x-axis position (**+** instead of **-**).

Step 4: We can now add in an x-axis, with tick labels at the default x-axis positions of 1–5 so that they will sit right in the middle of the two boxes at each level of watering. We do this using the **axis** function, used previously in chapter 4 (see also section 7.5.2 for further details on axis customization):

```
axis(side=1, at=1:5, labels=c(1:5))
```

Step 5: And finally, because we have plotted the data from the two different treatments alongside each other, we also need to include a legend to explain our use of different colours:

```
legend("topleft", bty="n", legend=c("Uncovered","Covered"),
fill=c("gray","yellow"))
```

In Figure 5.9, we now have data for the two treatments, '**uncovered**' as grey boxes and '**covered**' as yellow boxes, plotted alongside each other, with bigger gaps along the x-axis between the different levels of watering than between the two grouped treatments at any one level of watering. Of course, the arguments used to play with space here (primarily **boxwex** and **at**) are highly customizable, so feel free to play around with values and decide which help present your data as clearly as possible.

As a small disclaimer, ordinarily we would not recommend making boxplots for only four discrete values, as we have for each treatment in this experiment (see our advice in sections 4.1 and 5.1 on treating data with seven or fewer different values as qualitative data instead). However, here we wanted to demonstrate grouped multiple-samples plots using data from a clear and uncomplicated experiment, without an overwhelming volume of data.

> 💡 **Key point**
>
> Not only can boxplots handle multiple samples with ease, if those samples are grouped then that can be effectively represented in your boxplot too.

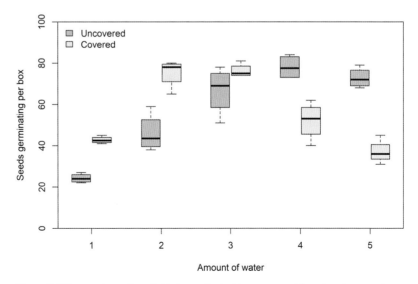

Figure 5.9 The number of seeds that germinated in uncovered (grey) and covered (yellow) boxes experiencing different levels of watering, where 1 represents the least amount of water used and 5 represents the most (n=4 for all except covered boxes with a watering level of 5, for which n=3). Adapted from data from Chatfield (1982).

5.7   Stretch your understanding

To help solidify and stretch your understanding of the ideas covered in this chapter, we have provided three data sets in the online resources for you to have a go at making histograms and boxplots with. For each data set we explain what we're hoping to explore through visualizing the data—once you've had a go yourself, check out how we approached this data in our online resource 'Authors' attempts for chapter 5' and associated 'R script for Authors' attempts chapter 5'.

The first example ('haemoglobin_and_elevation.csv') is adapted from a study that was interested in comparing the average haemoglobin concentration (g/dl) of American, Ethiopian, Tibetan, and Andean residents in order to better understand patterns of human adaptation to high-altitude hypoxia (Beall et al. 2002). The residents of the US lived at sea level, while those in Ethiopia, Tibet, and the Andes lived at heights of 3530 m, 4000 m, and 4000 m above sea level, respectively. Might a high-altitude lifestyle select for greater oxygen-carrying capabilities?

The second example ('nova_scotia_birds.csv') contains data from the National Audubon Society Christmas Bird Count (Audubon 2020). Since the beginning of the twentieth century there has been an annual bird count across Canada and the United States during the holidays, and the data set provided here contains the species counts from the years 2014–2018 from seven different survey areas within the Nova Scotia province of eastern Canada. We are interested in whether the total counts of species tend to vary noticeably between these different areas.

The third data set called 'caffeine.csv' shows finger tapping results from an experiment on the effect of caffeine consumption. Thirty college students were trained in finger tapping, where fingers are fitted with sensors and you have to tap them as many times as you can within a given time frame. They were then divided randomly into three groups and given different doses of caffeine (0, 100, and 200 mg). Two hours after treatment each participant was required to do finger tapping and the number of taps per minute was recorded (Draper and Smith 1981). The investigators then repeated this study with thirty professional gamers to find out if: i) gaming improves finger tapping abilities, and ii) if caffeine has more or less of an effect on finger tapping by professional gamers compared to college students. Think about all the different samples that will need to be presented here and whether there are any similarities between samples that would make grouping a sensible choice.

 Chapter Summary

- With multiple-samples data, we have two or more samples, with each sample from a different group of individuals. As with single-sample data, for each individual within the sample we have one single measurement that should either be continuous or discrete with quite a few different values.
- Multiple-samples boxplots are easy to read side by side, and provide a lot of detail about each sample.
- Histograms are great for showing the distribution of each sample, but it can be difficult to compare multiple histograms or interpret overlapping ones.

- Bar charts are generally less informative, giving less detail about the distribution of each sample, and so should only be used for displaying quantitative data after careful consideration.
- Grouped multiple-samples data is where different samples may logically be grouped together in a figure; boxplots are the best option here.

Online Resources

The following online resources are available for this chapter at www.oup.com/he/humphreys-obp1e:

- R script for chapter 5
- pig_litters.csv
- seed_data.csv
- Bar charts for quantitative multiple-samples data
- R script for Bar charts for quantitative multiple-samples data
- haemoglobin_and_elevation.csv
- nova_scotia_birds.csv
- caffeine.csv
- Authors' attempts for chapter 5
- R script for Authors' attempts chapter 5

Further Reading

- 'Bar Graphs Criticized for Misrepresenting Data':
  https://doi.org/10.1038/520589f
- 'Bar Graphs Depicting Averages are Perceptually Misinterpreted: The Within-the-bar Bias':
  https://doi.org/10.3758/s13423-012-0247-5
- 'Beyond Bar and Line Graphs: Time for a New Data Presentation Paradigm':
  https://doi.org/10.1371/journal.pbio.1002128
- 'Data Visualization' section of 'Transparent Reporting for Reproducible Science':
  https://doi.org/10.1002/jnr.23785

6 SCATTERPLOTS FOR QUANTITATIVE DATA

Learning objectives

By the end of this chapter you should be able to:

- Explain when a scatterplot is a good choice of data presentation.
- Produce clear and effective scatterplots.
- Present multiple samples of data, with separate lines showing trends in the data, on the same scatterplot.
- Understand when it is appropriate to use a univariate scatterplot (or 'strip-chart') to display each data point collected for a single variable individually.

6.1 Introduction: scatterplots and univariate scatterplots

Scatterplots are the most effective way of plotting data points where you have collected data on two variables and are interested in whether there is a relationship between those variables. In these cases, we have a sample of individuals, and for each individual we measure both traits. Scatterplots work well if these traits are continuous variables, or discrete quantitative variables with a logical ordering and at least seven different possible values, or a mix of the two.

When producing a scatterplot, we will plot one trait along the x-axis and one trait along the y-axis to get a sense of whether there is a relationship between the two. In essence, we're looking to see if there's any correlation between the two things we have measured. In Bigger Picture 6.1 we provide a fuller explanation of the meaning of correlations in science and what they can look like.

In this chapter, we will also cover when it is appropriate to present data as a univariate scatterplot (more commonly called a 'strip-chart'). Unlike the scatterplots that are the main focus of this chapter, strip-charts display all the individual data points collected for a single measured variable, instead of two measured variables. Strip-charts were mentioned in Scientific Approach 4.2 as an alternative to histograms or boxplots (or as an accompanying presentation tool), and here we will provide further advice on when it is useful to present each individual measurement of a quantitative data set instead of, or as well as, a figure (like a boxplot) that summarizes aspects of the data.

Bigger Picture 6.1
Scatterplot correlation

When scientists talk about a correlation between two variables, they mean a mutual relationship or connection between the values of the two things. You will find correlations discussed widely in statistical analyses as well as in graphical display of data, so it is worth taking the time to be clear on the variety of different forms a correlation might take. The various patterns of data shown in BP 6.1 Figure A should help you get to grips with what different correlations, varying in strength and direction, can look like.

Note: If you are interested in how we arranged the figure in BP 6.1 Figure A, see the online resource 'R script for chapter 6' and section 7.2.2 where we discuss multi-panel plots.

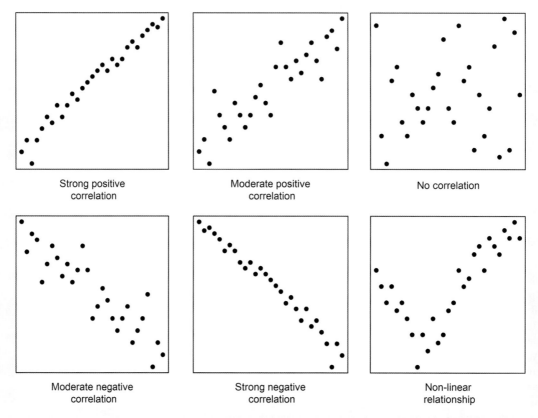

BP 6.1 Figure A Six different plots, each with the same number of data points (n=28), showing various patterns and the strengths and directions of some relationships that can be visualized by plotting data as a scatterplot.

 Key point

Scatterplots can be good for exploring whether, or how, two variables are related.

6.2   An example of continuous data

The example data set for this chapter is one of the built-in data sets that R comes with, called 'Puromycin'. To see the list of preloaded data sets R provides (as demo data for trying out different R code), you can simply run **data( )** through the R console. In this list, you will find 'Puromycin', described as 'Reaction Velocity of an Enzymatic Reaction'. To load and view this data run the following code:

```
data(Puromycin)
View(Puromycin)
```

To learn more about this data, try running the code:

```
?Puromycin
```

We can now see in the 'Help' tab of the bottom right-hand window of RStudio that this data comes from a study into the velocity of an enzymatic reaction involving 'untreated' cells or cells 'treated' with puromycin (Treloar 1974). Puromycin is an antibiotic that inhibits protein synthesis. The 'conc' column refers to the concentration of the substrate provided (mM). The 'rate' column details the initial rate (or velocity) of the reaction (disintegrations per min/min, or DPM/min), calculated as (counts/min)/min, where the number of counts per minute of radioactive product from the reaction was measured as a function of substrate concentration in parts per million (ppm). We are interested in whether there is a relationship between the substrate concentration and the rate of enzymatic reactions, and whether there is any difference in this relationship between the untreated cells and the cells treated with puromycin. We also provide the data as a .csv file in the online resources ('puromycin_R_data.csv') if you prefer to download this and load it into R with the name '**Puromycin**' (see the 'R Basics' guide from the online resources for chapter 1 for how to do this).

 Key point

Always take some time to fully understand how your data is organized before trying to ask R to perform any operations on it.

6.3   Looking for correlations with scatterplots

6.3.1   Simple scatterplot

Step 1: Once the data set is loaded, we might want to start by assigning short names to the different variables, for easy reference later on:

```
state<-Puromycin$state
conc<-Puromycin$conc
rate<-Puromycin$rate
```

We introduced this trick in section 4.2 but, as a reminder, in R variables can be called directly from a data set by naming the data set, followed by a $, then the name of the data column of interest with no spaces in between. The longer data-$-column format could be used throughout all graphing, but here we have decided to 'rename' the long format way of pulling out variables from the data set by simply telling R to assign these variables shorter format names (here the column names used in the data set).

Step 2: To start with, we can look to see if there is any correlation between substrate concentration and enzymatic reaction rate, regardless of whether cells have been treated with puromycin or not. You may have an idea of what this might be already, but let's see how visualizing these variables with a scatterplot can help confirm, question, or refine our original suspicions. First, we should have a quick look at the minimum and maximum values for both concentration and rate to see where we might want to set our axis limits:

```
summary(conc)
summary(rate)
```

The **summary** function returns six different descriptive statistics, but what we are interested in here are the values given under 'Min.' and 'Max.' (see Bigger Picture 4.2 for definitions of the other outputs). Substrate concentration ranges from 0.02 to 1.10 mM, while reaction rate ranges from 47 to 207 DPM/min.

Step 3: The code for a basic scatterplot is now really easy: we simply use **plot(** and then state which variable we want on the x-axis, followed by which variable we want on the y-axis. If you are unsure of how to decide which variable should be on the x-axis and which should be on the y-axis, this is discussed in Bigger Picture 6.2. Bearing in mind the minimum and maximum values, we can also set reasonable axis limits here:

```
plot(conc,rate, xlim=c(0,1.2), ylim=c(0,250), xaxs = "i",
yaxs = "i", xlab="Concentration (mM)", ylab="Enzymatic
reaction rate (DPM/min)")
```

In Figure 6.1, it looks as though there is a moderate positive correlation (see Bigger Picture 6.1 for a recap of correlations) between substrate concentration and the enzymatic reaction rate; as substrate concentration increases so does the enzymatic reaction rate. However, the relationship is more complex than a simple positive linear correlation, since the initial rising trend at low concentration values seems to flatten at higher values.

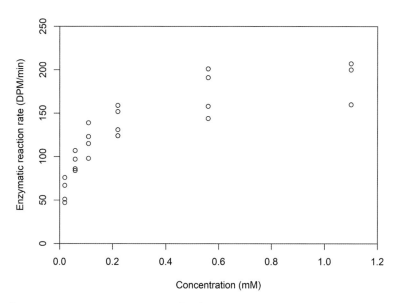

Figure 6.1 As substrate concentration (mM) increases so does enzymatic reaction rate (disintegrations per min/min, or DPM/min), at least up to around 0.6 mM. Data from Treloar (1974).

Bigger Picture 6.2
Which variable should go on which axis?

For many graph types, deciding which variable should be on the x-axis and which variable should be on the y-axis requires a bit of thinking about the study from which the data was sourced. Below are four scenarios you can use to help you with this thinking:

- If you were to conduct an experiment where you have control over one variable and you are measuring a second uncontrolled variable, the controlled variable is the independent variable and the uncontrolled variable is the dependent variable. You are measuring any change in the dependent variable in response to changes (determined by you) in the independent variable. Independent variables are almost always on the x-axis, and dependent variables are almost always on the y-axis.

- If you were to conduct a different study, where you were simply observing two variables but one variable is thought—perhaps due to past research—to be 'explanatory' (or the 'cause') and another variable is seen as being 'explained' (or the 'effect'), the convention is to put the explanatory variable on the x-axis and the variable being explained by it on the y-axis.

- If you were to conduct a third study, where this time you want to make predictions about one variable based on the other, the variable you want to predict should be put on the y-axis and the variable you want to base the predictions on should be put on the x-axis.

- If you were to conduct a fourth study, where you are interested in whether there is a relationship between two variables but there is no predicted dependency between the variables (no cause-and-effect relationship) and you are not making predictions about one variable based on the other, then either variable could go on the x-axis (as long as both axes are clearly labelled).

Step 4: We can make the precise values indicated by the scatterplot points easier to read by adding grid lines to this plot. Horizontal grid lines (see sections 3.2.2 and 4.3.5 for examples) and vertical grid lines (see section 3.5.1) have already been covered earlier in the book, and here we will again use `par(new=TRUE)`. First, we need to make a blank plot (but with our x- and y-axis set as needed) by substituting our variables with -5 (an arbitrary value, but importantly one that will not appear within the axis limits we define) and setting our axes with blank labels and no tick marks:

```
plot(-5, xlim=c(0,1.2), ylim=c(0,250), xaxs = "i",yaxs = "i",
xlab="", ylab="",xaxt="n",yaxt="n")
```

Step 5: Because we have two continuous variables here, we will now plot both vertical (denoted with **v**) and horizontal (denoted with **h**) grid lines. We differentiate major and minor grid lines by using different colours, but remember you could also (or alternatively) adjust line thickness using `lwd=`:

```
abline(h=(seq(0,250,10)), col="lightgray", lty=1)
abline(h=(seq(0,250,50)), col="darkgray", lty=1)
abline(v=(seq(0,1.2,0.05)), col="lightgray", lty=1)
abline(v=(seq(0,1.2,0.2)), col="darkgray", lty=1)
```

Step 6: Now we run the command `par(new=TRUE)`, followed by a full line of code for the scatterplot we want to position on top of the existing grid lines:

```
par(new=TRUE)
plot(conc,rate, xlim=c(0,1.2), ylim=c(0,250), xaxs = "i",yaxs = "i",
    xlab="Concentration (mM)", ylab="Enzymatic reaction rate (DPM/min)",
    col="purple", pch=24, bg="yellow", cex=1.5,lwd=2)
```

As ever when grid lines are used, it is important that the limits set for our axes match those defined in the **abline** code for the grid lines so that the scatterplot is perfectly overlaid. We also include in this final plot details such as axis labels and we allow tick marks to be added automatically. You may also notice that we have added code to modify the appearance of the data points so that they look clearer on top of the grid lines. R has loads of ways we can customize the shape, size, and colour of the points we plot; in Scientific Approach 6.1 we detail ways to modify points. **pch** tells R what shape point we want, **bg** tells it what colour we want it filled with (though note this only applies to certain shapes, as is also mentioned in Scientific Approach 6.1), **col** defines the outline colour, **cex** is the shape size, and **lwd** is the width of the outline.

Scientific Approach 6.1
Modifying points

Different scientists will have different aesthetic preferences for how they like the points in their scatterplots (or other graphs) to appear, and this might vary depending on the data being presented. As usual, R has the flexibility to accommodate a very broad range of approaches to the design of points.

The range of point shapes available in R, illustrated in SA 6.1 Figure A, can be 'called' with their corresponding numbers using **pch**.

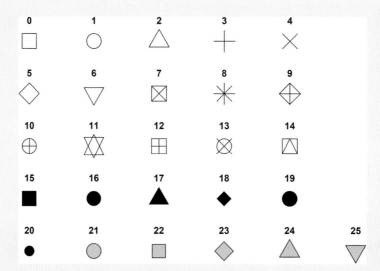

SA 6.1 Figure A The various point shapes that can be used in R (modified from http://www.sthda.com/english/wiki/r-plot-pch-symbols-the-different-point-shapes-available-in-r).

The colour and size of the points shown in SA 6.1 Figure A can further be modified with:

- **col**: colour (code or name) to use for the points

- **bg**: the background (or fill) colour for the open plot symbols ('**skyblue**' here). It can be used only when **pch** = 21 to 25.

- **cex**: the size of **pch** symbols. The larger the value of **cex**, the larger you make the character (generally values between 0.5 and 3 work best).

- **lwd**: the line width for the plotting symbols.

Step 7: As a final flourish to our scatterplot, we might want to plot a line that summarizes the major trend in the data to help us determine whether there is any correlation between substrate concentration and enzymatic reaction rate. There are lots of statistical approaches that can be used to derive such a line. We will use a simple way to derive a straight line that we will call the 'line of best fit'. We will here plot a straight line of best fit, but see section 6.3.3 for variations on this. Run the command:

```
abline(lm(rate~conc))
```

This plots the linear model (**lm**) of rate against concentration. In Bigger Picture 6.3 we explain linear models and how to interpret their outputs in R. Make sure you order the linear model code so that your y-axis data is first and then your x-axis: here we are seeing if rate depends on (~) concentration of substrate.

The line of best fit can be edited (changing colour [**col**], line type [**lty**], and line width [**lwd**]) using arguments discussed further in section 7.3.4. We might not always want our line of best fit to run through the y-axis. In cases where it is unwise to extrapolate trends beyond the data collected, we can instead plot a more restricted line of best fit, which we discuss further in Scientific Approach 6.2.

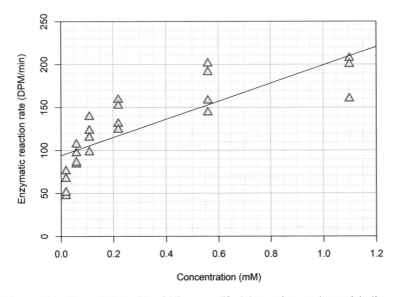

Figure 6.2 As Figure 6.1 but with grid lines, modified data points, and a straight line of best fit that suggests a positive correlation between concentration of substrate and rate of reaction.

Bigger Picture 6.3
Explaining linear models

Linear models are fundamental components of most introductory statistics courses and texts you will experience. To help you see how graphing fits into the bigger picture of statistical exploration of data, we will explore them very briefly here. A linear model describes the general linear tendency of the data, and therefore can be plotted as a straight line that best fits your data. Looking at the linear model used in the puromycin example itself, we simply run the R command:

```
lm(rate~conc)
```

which yields the output:

```
Call:
lm(formula = rate ~ conc)
Coefficients:
  (Intercept) conc
     93.92   105.40
```

This tells us that the straight line of best fit is:

```
Rate = 93.92 + 105.4Conc
```

At the top of the output, R just tells us the formula we are looking at is a linear model. The first number under 'Coefficients' is then the y-intercept, which can be thought of as the value for rate of a (notional) reaction with a vanishingly small concentration of substrate (or where your line of best fit is going to 'pass through' the y-axis). The second number is the gradient (or slope of your line). This value suggests that every 1 DPM/min increase in rate is on average associated with a 105.4 increase in concentration (mM). We know it is an increase because the value is positive. This fits with our general intuition that as the concentration of substrate increases, rate of reaction increases. **abline** can be used to plot this linear model for us, allowing us to visualize the upward-sloping trendline.

Scientific Approach 6.2
Restricting lines of best fit

A broad approach in data analysis is to fit a model (often something as simple as a straight line) to your data. However, it is dangerous to extrapolate your model beyond the bounds of the data. That is, it is safest for you to think of your model as only being reliable when it sits within the range of values for which you actually have data. For straight line models, we might want to remind ourselves of this by not having our fitted straight line run across the full length of our x-axis, but only where we feel more comfortable with the model's validity. R lets us restrict the length of our line to achieve this, as we will demonstrate with our **Puromycin** data. First we need to check what range of x-axis values we want the line to run through:

```
summary(conc)
```

The minimum concentration is 0.02 mM, and the maximum is 1.1 mM. We tell R these values of where we want the line to start (**x0**) and end (**x1**):

```
x0<-0.02
x1<-1.1
```

In Bigger Picture 6.3 we looked at the model **lm(rate~conc)**, which told us that the line of best fit is:

```
Rate = 93.92 + 105.4Conc
```

We tell R these values too:

```
a<-93.92
b<-105.4
```

Now we simply plot our graph as before and add our line using the **segments function** in the following command that can be easily copied and pasted:

```
segments(x0, a+b*x0, x1, a+b*x1, col =
"seagreen3", lty = 5, lwd=3)
```

As with other line functions, the segments command can be modified with your preferred colour (**col**, here 'seagreen3'), line type (**lty**, here 5, or 'longdash') and line width (**lwd**), as discussed further in section 7.3.4. The code here gives us SA 6.2 Figure A.

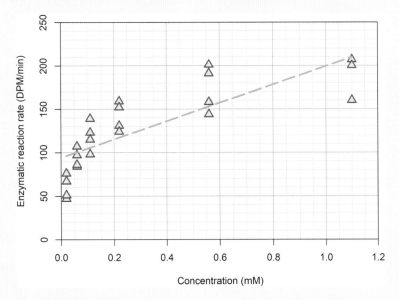

SA 6.2 Figure A A scatterplot illustrating the linear relationship between substrate concentration and enzymatic reaction rate with a restricted line of best fit. Data from Treloar (1974).

In Figure 6.2, our straight line of best fit suggests that there is a fairly strong positive correlation between the concentration of substrate and the enzymatic rate of reaction. This is perhaps not very surprising, but remember that the puromycin data set constitutes data on untreated cells and cells treated with puromycin. Next, we'll have a look at how to differentiate these data points in our scatterplot and see whether there is any difference in the correlation between the two cell states.

6.3.2 Scatterplot with multiple samples

We have seen in section 6.3.1 that, generally, reaction rate seems to increase with substrate concentration. Now we turn our attention to whether substrate concentration affects reaction rate differently depending on whether cells are untreated or treated with puromycin. To do this we can subset our data. You may be familiar with subsetting from section 5.6.2, where we produced a grouped boxplot with data from multiple samples; we use exactly the same method here.

Step 1: To start with, we can have a quick look at a summary of the different cell states in the data. This is more necessary where you start with an unknown number of samples in a given variable, but is still a useful way to check that your data are as expected:

```
summary(state)
```

Here we have only two different states of cells: treated (of which there were 12 recorded observations) and untreated (of which there were 11 recorded observations).

Step 2: We can now tell R to consider these two cell states each as mini data sets in their own right with the **subset** function:

```
treated <- subset(Puromycin, state == "treated")
untreated <- subset(Puromycin, state == "untreated")
```

As a brief reminder, 'subsetting' is useful for situations when you just want to use a particular part of your data set for plotting or carrying out statistical tests. Here, the function tells R we want to **subset** data from the puromycin data set where the 'state' variable of that data set equals (== in R) 'treated' (or 'untreated') and add it to a new data group called '**treated**' (or '**untreated**') named in front of the **<-** symbol.

Step 3: Next, we plot our grid lines exactly as we did before:

```
plot(-5, xlim=c(0,1.2), ylim=c(0,250), xaxs = "i",yaxs = "i",
  xlab="", ylab="",xaxt="n",yaxt="n")
abline(h=(seq(0,250,10)), col="lightgray", lty=1)
abline(h=(seq(0,250,50)), col="darkgray", lty=1)
abline(v=(seq(0,1.2,0.05)), col="lightgray", lty=1)
abline(v=(seq(0,1.2,0.2)), col="darkgray", lty=1)
```

Step 4: Now, when we plot our data points onto the grid we will plot the two subsetted data sets onto the same axes separately, establishing different colours and point shapes for each. A key point to remember here is to use the full names for each variable; including the name of the subsetted data set we want, then the $ sign, then the variable. Also, remember to use the command **par(new=TRUE)** before adding each set of data to the same plot:

```
par(new=TRUE)
plot(treated$conc,treated$rate, xlim=c(0,1.2), ylim=c(0,250),
  xaxs = "i",yaxs = "i", xlab="Concentration (mM)", ylab=
  "Enzymatic reaction rate (DPM/min)", pch=15, col="orange")
par(new=TRUE)
plot(untreated$conc,untreated$rate, xlim=c(0,1.2),
  ylim=c(0,250), xaxs = "i",yaxs = "i",
  xlab="", ylab="", xaxt="n",yaxt="n",
  pch=19, col="blue")
```

Note that for the second data set we plotted we do not need to include axis labels (**xlab=""**, **ylab=""**) or axis tick labels (**xaxt="n",yaxt="n"**) a second time, as they have already been added with the first data set's code.

Step 5: In the same way, it is now easy to add colour-matching lines of best fit for both cell states too:

```
abline(lm(treated$rate~treated$conc),col="orange",lwd=2)
abline(lm(untreated$rate~untreated$conc),col="blue",lwd=2)
```

Step 6: Now, in order to explain our use of colour and symbols, we finish by adding a legend. Here we have specified x- and y-coordinates for its placement (rather than using 'bottomright'), and we have made the background white so that the legend is easy to read on top of the grid lines:

```
legend(0.9,70, bg="white", pch = c(15,19), col = c("orange",
"blue"), legend=c("Treated","Untreated"))
```

We now have a very neat scatterplot, that clearly distinguishes and labels the data from the cells treated with puromycin and the untreated cells (take a look at Figure 6.3). Looking at the two straight lines of best fit, both cell states seem to show a positive correlation between substrate concentration and enzymatic re-action rate, although the treated cells seem to reach higher reaction rates across all concentrations of substrate. However, looking closely at the data points now we have differentiated the cell states, it does not look like the relationship be-tween substrate concentration and reaction rate is necessarily a straight, linear one. For both cell states, there appears to be an initial steep increase in reaction rate as concentration initially increases from 0.02 mM to 0.22 mM, which then reduces in gradient from between 0.22 mM and 0.56 mM, so that it is almost level from 0.56 mM to 1.1 mM. For this reason, we will next demonstrate how to fit curved lines of best fit to better follow these kinds of relationship.

6.3.3 Refined scatterplot: adding curved lines of best fit, labelling individual data points, reference lines, and text

In order to fit curved lines of best fit to the data, we will use the lowess func-tion. There are lots of other approaches (see Further Reading), and some will be more appropriate than lowess for particular situations. We only highlight this function here because it is easy to use and works widely. However, no matter how you go about generating lines that represent the main features

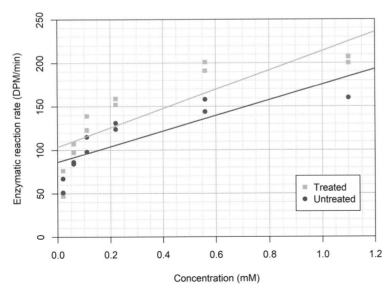

Figure 6.3 Both cells treated with puromycin and untreated cells appear to show a positive correlation between substrate concentration and enzymatic reaction rate. Data from Treloar (1974).

of your data, the principles that we introduce in this chapter should help you to display that line effectively alongside your data. The method LOWESS (locally weighted scatterplot smoothing) is designed for adding smooth curves to scatterplots with a single predictor variable. You might also come across the LOESS (locally estimated scatterplot smoothing) method, which is a generalized version of LOWESS that is also commonly used to fit curves to data (via the `lowess` function in R). But LOESS models can include multiple predictor variables, factors, and interactions, and so their implementation is more complicated. For this reason, we will stick to the LOWESS method, but see the Further Reading at the end of this chapter for details on how both methods work and the difference between the two.

Step 1: Before we add new, curved lines of best fit to the data, you can prepare the plot in exactly the same way as before. Simply run the code for the previous figure up to just before we added the straight lines of best fit, and then also run the commands for the legend.

Step 2: Now, for each of our subsets ('`treated`' and '`untreated`') we will create a list of ordered x-coordinates and smoothed y-values using the `lowess` function. The basic syntax for `lowess` is simply `lowess(x,y)`—we have one predictor (x) and one outcome variable (y)—so here we have:

```
treatedcurve<-lowess(treated$conc,treated$rate)
untreatedcurve<-lowess(untreated$conc,untreated$rate)
```

If you run either '`treatedcurve`' or '`untreatedcurve`' through the Console you can see how the values for the curve are organized.

Step 3: We will now use these lists of values to draw a curved line for each cell state on our scatterplot. To do this, we use the `lines` function in R (see section 7.3.4 for further line customization). `lines` can make sense of coordinates presented in various different ways, but it essentially works by joining the corresponding points it is given with line segments. As ever, we can also specify the colour and width of the lines:

```
lines(treatedcurve,col="orange",lwd=2)
lines(untreatedcurve,col="blue",lwd=2)
```

This leaves us with a scatterplot that has two clear, distinguished curved lines of best fit that match the trend of both our subsets well (take a look at Figure 6.4). From what we know of how enzymes function, it makes sense that reaction rate could not continue to increase in a linear fashion despite increasing substrate concentration—a maximum reaction rate threshold will be reached, above which increasing substrate concentration will not increase the reaction rate further. Here, the default smoothing by the `lowess` function was a pretty good fit for our data, but in Scientific Approach 6.3 we explain how to edit the smoothing parameter to suit your needs.

Imagine now that we already have data on another drug that appears to influence the rate of enzymatic reactions. The other drug has only been tested at substrate concentrations of 0.56 mM, but it increased the rate of enzymatic reactions in treated cells relative to untreated cells by 40 DPM/min. We might now want to use our figure to clearly communicate whether puromycin appears to cause a greater or lesser increase in enzymatic reaction compared to the other drug. We are interested in comparing specific reaction rates between the treated and untreated cells in order to compare their relative performance at a particular substrate concentration.

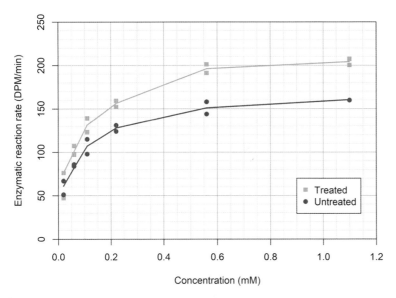

Figure 6.4 With curved lines of best fit, it is clearer to see the trends for both cell states. After an initial steep increase in reaction rate as concentration increases from 0.02 mM to 0.22 mM, the increase in reaction rate with increasing concentration slows between 0.22 mM and 0.56 mM, and seems to be levelling out towards a maximum reaction rate between 0.56 mM to 1.1 mM (Treloar 1974).

Scientific Approach 6.3
Smoothing with **lowess**

Different scientists might take different approaches as to how much smoothing should be applied to a fitted model to best characterize the features of the data of interest to them. As you would expect, R offers the flexibility to accommodate this range of different approaches. To achieve different levels of 'smoothness' when fitting a curved line to a scatterplot, you simply need to adjust the smoother span parameter 'f' in the **lowess** function. 'f' is defined as the proportion of data points in the plot which influence the smoothing at each value, so the larger the value assigned to **f**, the more extreme the smoothing of your line. By default, the **lowess** function applies a smoother span parameter of 2/3, but this can be changed by adding in your own value for **f** when using this function.

In the example given in SA 6.3 Figure A, we used another of R's built-in data sets (Theoph) to examine the pharmacokinetics of theophylline in three subjects (1, 2,

and 6) over the following day after drug administration. The three panels show (in order from top to bottom) the lines with the smoother span parameter (**f**) set to 0.1, 0.66 (the default for **lowess**), and 1. In terms of code syntax, for subject 1 the lists of values for these curves were prepared as follows:

Less smoothing:

```
sub1curve0.1<-lowess(sub1$Time,sub1$conc,
f=0.1)
```

Default:

```
sub1curve<-lowess(sub1$Time,sub1$conc)
```

More smoothing:

```
sub1curve1<-lowess(sub1$Time,sub1$conc,
f=1)
```

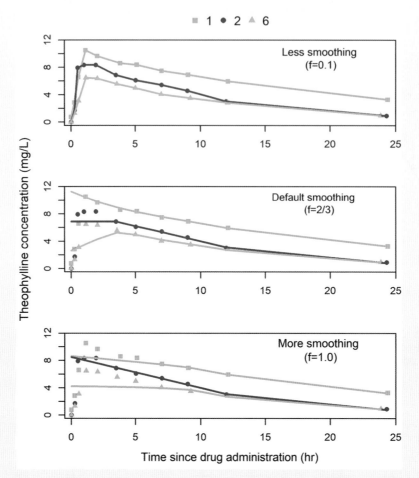

SA 6.3 Figure A Each plot shows the same data on the change in theophylline concentration (mg/L) in three subjects (1, 2, and 6) over the following day after drug administration, but the smoothness parameter used when drawing the lines increases as you move down the plots. Data reported by Boeckmann et al. (1994).

The question then remains, how much smoothing should you apply to your data? This decision depends on how much resolution you want to view the trend of your data with. Adding highly smoothed lines can help detect trends in the presence of very noisy data where the shape of the trend would be otherwise unknown—in these cases you probably would not want fine detail of the relative ups and downs of short spans of the y-axis. However, when you have data that seems to follow a general trend, lines with relatively less smoothing can capture more of the details within shorter spans of the y-axis without creating very 'wobbly' and confusing trend lines. Essentially, how much smoothing should be determined by how much detail about the data variation you want to include across the span of the graph, and whether you are mainly interested in a very general overall trend. In SA 6.3 Figure A, for example, we think relatively little smoothing is most useful in helping to capture the detail of the initial peak in theophylline concentration across all subjects, ahead of the steady decline over time.

Note: If you are interested in how we arranged the figure in SA 6.3 Figure A, see the online resource 'R script for chapter 6' and section 7.2.2 where we specifically discuss multi-panel plots.

To achieve this, we first might want to present the specific values our data points represent on our figure (though for this we would generally encourage the use of a table, as we discuss in section 2.3). We can then think about adding reference lines and text to make our findings clear.

Step 4: R has an easy way you can label scatterplot points but, as you might imagine, extra labels would look pretty messy on top of grid lines and around lines of best fit. So, first produce a figure exactly like Figure 6.4, but without the underlying grid lines.

Step 5: Now, using the tidied variable names we defined earlier on, we can run a simple command using the **text** function to quickly label all the data points:

```
text(rate~conc, labels=rate,pos=2, cex= 0.8,font=1)
```

labels tells R what variables we want the point labels to come from (the reaction rates in this instance), and **pos** refers to the position of the text relative to the data points (values of 1, 2, 3, and 4 respectively indicate positions below, to the left of, above, and to the right). We have also selected **font** face 1, which is normal text and would be the default anyway in R—however, we include this for context as we specify a bold font shortly. Notice we don't specify any subsets here (no $ symbols) so the text labels for rate~conc will contain all the rates for both treated and untreated cells. See section 7.3.5 for further details on adding text to figures.

Note: The labels assigned to data points do not have to be either of the variables in the x~y formula shown in your figure, and do not have to be numerical either. They could come from an entirely different variable in your data set, as long as each data point can be associated with a label from that variable, e.g. you might wish to show names of individuals or countries.

In Figure 6.5, unfortunately the labelling of individual data points looks pretty messy—this would not really change even if we reposition the labels

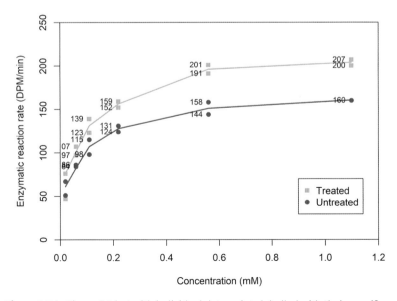

Figure 6.5 As Figure 6.4 but with individual data points labelled with their specific reaction rates and no grid lines.

relative to the points (using **pos**). Where data points are close together, the labels overlap and are rather illegible. For some sets of data where the points are less clustered, or where it is particularly useful for individuals to be identifiable, then labelling individual data points might be helpful. Generally though, as mentioned earlier, if you want to present all of the numerical values of a data set, a table is a clearer format to use (as we suggest in section 2.3). Going back to our hypothetical scenario, the drug we want to compare puromycin to has only been tested at 0.56 mM substrate concentration. The values for the untreated and treated cells at 0.56 mM in our puromycin experiment are reasonably clear, so let us now produce that figure again but this time keeping only those data point labels at the concentration of interest to us.

Step 6: Once we have reproduced the figure without any data labels, we can use subsetting to separate out our data values of interest. Here, we tell R to subset all the rows of data where the concentration is 0.56 mM into the new grouping 'interestvals':

```
interestvals<-subset(Puromycin,conc==0.56)
```

Step 7: Now we have the new subset, we can just do as we did before, but this time specifying the subset 'interestvals'. We can also try listing different positions for the labels in order to space them out a little more—here alternating to the right (**pos**=4) and to the left (**pos**=2) of the data points:

```
text(interestvals$rate~interestvals$conc, labels=
interestvals$rate,pos=c(4,2,4,2), cex= 0.8,font=1)
```

Step 8: We can also highlight that these reaction rates are specific to the substrate concentration of 0.56 mM by adding a reference line and associated text there.

```
abline(v=0.56,lwd=2, lty=2,col="purple")
text(0.65,50, labels="0.56mM",col="purple",font=2)
```

In this use of the **text** function, we specify the x- and y-coordinates for the placement, before stating what we want printed with the **labels** argument. We also specify a bold font (**font=2**), rather than the default font face (**font=1**) used previously in this section; font style is discussed more in section 7.3.2.

Step 9: Finally, it might be helpful for us to present the mean reaction rates of the cells treated with puromycin and the untreated cells at 0.56 mM substrate concentration to aid our comparison. This time when we use the **text** function again, we also get R to calculate the means of the values we have presented on our figure. You might recall our use of the **paste** function in section 2.2.1.2 to print values alongside % symbols as pie chart segment labels. See section 7.3.5 where we further explain the use of **text** and **paste** in the creation of Figure 6.6, and see section 8.6 and Scientific Approach 8.1 for even more on the **paste** function (and its use alongside the **expression** function).

Run the commands:

```
text(0.7,220, labels=paste("mean =",mean(c(191,201))),
  col="orange",font=2)
text(0.7,125, labels=paste("mean =",mean(c(144,158))),
  col="blue",font=2)
```

Looking at the resulting Figure 6.6, we can now easily compare the mean enzymatic reaction rates and, if desired, the specific reaction rates of individual samples of cells treated with puromycin and untreated cells when substrate concentration is 0.56mM. The mean enzymatic reaction rate of cells treated with

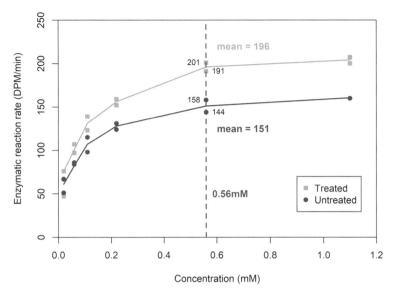

Figure 6.6 A high-quality scatterplot showing how the enzymatic reaction rate changes with substrate concentration for cells treated with puromycin and untreated cells (data from Treloar 1974). At substrate concentration 0.56 mM, the mean enzymatic reaction rate of treated cells is 45 DPM/min greater than untreated cells.

puromycin is 45 DPM/min greater than untreated cells. Thinking back to our other drug, which increased the rate of enzymatic reactions in treated cells relative to untreated cells by 40 DPM/min, we might suggest that puromycin has more of an effect on reaction rates than the other drug. But we are also good scientists! We know that a sample size of two for each condition in our puromycin study is certainly not enough for us to draw any solid conclusions about its effectiveness relative to another drug. This is particularly true given that in Figure 6.6 we can see that the difference between the maximum reaction rate of untreated cells (158 DPM/min) and the minimum reaction rate of puromycin-treated cells (191 DPM/min) at 0.56 mM is only 33 DPM/min. Nevertheless, we have demonstrated how scatterplots can be used to neatly present whole data sets and customized for the purpose of highlighting particular findings.

 Key point

You can convey a lot of information in a single scatterplot—you can use different colours or shapes of points to convey information beyond the values of the variables on the two axes, and you can fit lines of best fit through the data to help emphasize general trends.

6.4   Presenting time series

6.4.1   Refined time series

In the scatterplots covered in this chapter already, we have been interested in fitting a line of best fit to show the general trend across our data. However, you may sometimes have a data set where you want to join together the data points with a line. This could be particularly useful when illustrating change in one continuous variable over time (another continuous variable)—a time series. We will look at how to connect the points of a scatterplot showing change over time, using data on global land and ocean temperature anomalies with respect to the twentieth-century average ('land_and_ocean.csv') (NOAA National Centers for Environmental Information 2020).

Step 1: First, as ever, we must load in and view the data (see the 'R Basics' guide from the online resources for chapter 1 if you are unsure how to input data from a .csv file)—we name the data set '**temps**'. We can then also look at summaries of the variables we'll be plotting.

```
summary(temps$year)
summary(temps$value)
```

This simple data set has only two columns: 'year', which is self-explanatory, and 'value', which is the global land and ocean temperature anomaly for that year in degrees Celsius. When we look at summaries of the data, we can see that we have records of data running from 1880 all the way to 2019, and that our temperature anomaly values span negative as well as positive values. In order to demonstrate more clearly whether there is a consistent trend in temperatures over recent decades—and to demonstrate subsetting again—let's differentiate negative and positive values with different colours.

Step 2: As we did in section 6.3.3, we need here to subset our data based on the value of a variable. The following code groups rows of data where the temperature anomaly ('**value**') is less than or equal to (<=) 0 as '**neg**' and values greater than (>) 0 as '**pos**':

```
neg<-subset(temps,value<=0)
pos<-subset(temps,value>0)
```

Step 3: Now, using the minimum and maximum x- and y-values we saw using the **summary** function above, we can prepare suitable horizontal grid lines (just as with a normal scatterplot). We also here chose to add a thicker, darker line to mark 0 degrees Celsius and clearly separate the negative and positive values:

```
plot(400, xlim=c(1880,2020), ylim=c(-0.6,1.2), xaxs = "i",yaxs = "i",
  xlab="Year", ylab=expression(paste("Temperature Anomaly
  ( ",degree,"C)")))
abline(h=(seq(-0.6,1.2,0.2)), col="lightgray", lty=1)
abline(h=0, col="darkgray", lwd=2.5)
```

Note our use of the **expression** and **paste** functions to add in the mathematical symbol for degree (see section 8.6 and Scientific Approach 8.1 for details and further examples of multicomponent **expression** commands).

Step 4: We can now easily produce a scatterplot of both our subsets of data, differentiating by colour between negative and positive values, using code we have seen before (and remembering to use **par(new=TRUE)**):

```
par(new=TRUE)
plot(neg$year,neg$value, xlim=c(1880,2020), ylim=c(-0.6,1.2),
  xaxs = "i",yaxs = "i", xlab="", ylab="", xaxt="n",yaxt="n",
  pch=16, col="darkblue")
par(new=TRUE)
plot(pos$year,pos$value, xlim=c(1880,2020), ylim=c(-0.6,1.2),
  xaxs = "i",yaxs = "i", xlab="", ylab="", xaxt="n",yaxt="n",
  pch=16, col="red")
```

Step 5: Here comes the new code! The key step for joining data points together is this: ordering data points in terms of the x-axis values increasing (here time, because we are plotting a time series). We do this by using the **order** function to sort our data set by year. Although we split our data into two subsets (to differentiate negative and positive values), we want our line to connect all the data points because they all represent global land and sea temperature anomalies and we want to show the year-to-year changes across the entire time period. Therefore, when we use the **order** function in the following code, we order the data rows of the entire original '**temps**' data set by '**year**' and save this as the '**sortedvals**' data set:

```
sortedvals <- temps[order(temps$year),]
```

This command may look a little complex, but the syntax can be picked apart for use in other situations when you want to organize a data set by the values of one of the variables in that data set. After stating a new name for the ordered data, followed by the **<-** assignment operator, the name of the data set that needs reordering should then be followed by square brackets. Within the square brackets, the **order** function itself is positioned ahead of the bracketed variable that will determine the order for the whole data set; this is followed by a comma.

If you had multiple samples that showed change in an x-variable over time, you could add separate time series lines to each by using **order** to rearrange subsets of the different samples.

Step 6: We now easily join together our data points by using the **lines** function to plot from our reordered data '**sortedvals**'.

```
lines(sortedvals$year,sortedvals$value,lty=1,lwd=1,col="purple")
```

The resulting time series (reproduced in Figure 6.7) allows us to see the year-to-year variation in land and ocean temperature anomalies, through the peaks and

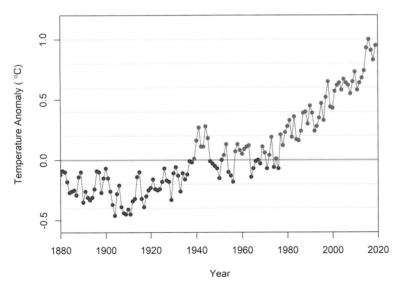

Figure 6.7 A high-quality time series showing change in global land and ocean temperature anomalies (with respect to the twentieth-century average) from 1880 to 2019 (NOAA National Centers for Environmental Information 2020), with negative or 0°C temperature anomalies in blue and positive temperature anomalies in red.

troughs of the connecting lines, but our use of colour also allows us to clearly see the marked increase in temperature anomalies that has steadily grown since around 1980.

We do need to confess here that ordering the data was not strictly necessary for the data set used in this example. This is because the .csv file we provided already had the values presented in the sequence of increasing years. However, it is a good (and not at all time-consuming) habit to get into, so that when you are faced with unordered data you know what to do.

6.4.2 The importance of order in a time series

To demonstrate the importance of ordering data when plotting a time series, we have adapted another of R's preloaded data sets: 'lh'. 'lh' details the concentration of luteinizing hormone (LH) in blood samples at 10-minute intervals from a human female (Diggle 1990: table A.1, series 3), but the version of the data we provide in the online resources as a .csv file ('LH_blood.csv') has the order of these intervals and the corresponding concentrations of LH in the blood samples totally rearranged. Have a look at this data set (and the code we provide) yourself—see if you can guess which panel from Figure 6.8 below (a or b) has had the order of its time series corrected!

> ### 💡 Key point
>
> A very common variable to show on the x-axis of a scatterplot is time, progressing from left to right along the axis.

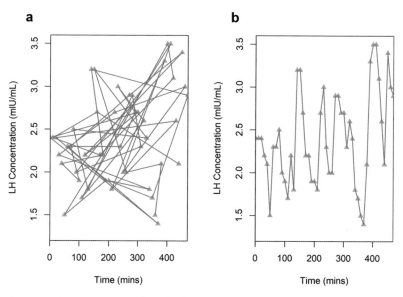

Figure 6.8 The concentration (mIU/mL) of luteinizing hormone (LH) in blood samples from a human female at 10-minute intervals: a) shows the output when the command using the **lines** function is run without the values being ordered by time, b) shows the output when the command using the **lines** function is run after values have been sorted by time.

6.5   Univariate scatterplots (strip-charts)

In Scientific Approach 4.2, we mentioned strip-charts as a possible additional feature for, or alternative to, boxplots. Strip-charts are essentially scatterplots for a single measured variable (hence, 'univariate scatterplots'). As they present all of the individual data points for a single quantitative variable along one axis, you can imagine that things will get messy quickly with more than around 20 data points. Of course, the spread of these points will influence the clarity of your findings, and strip-charts might be useful on occasions where you want to highlight a bimodal or other multimodal distribution in your data. However, where you have lots of data points, a violin plot may still be a better option for clear presentation of the uneven spread of data (see the Further Reading in chapter 4 for useful links). To our minds, strip-charts are typically a better alternative to a boxplot or a histogram only when sample size is low but there is still a benefit to presenting the data visually rather than in table form.

Here, we will briefly demonstrate how to use R's **stripchart** function using a data set with small sample sizes. If you load and view the 'birds_and_pterosaurs.csv' file from the online resources (we name the data '**flight**'), you will find data on the masses (kg) of six extinct pterosaur individuals and 10 extant seabird individuals (adapted from Witton and Habib 2010). There has been some debate over recent years about the flight capabilities of pterosaurs. We are interested in comparing the estimated masses of some of these extinct flying reptiles with the masses of some of the seabirds of today (including albatrosses, which are among the largest extant flying birds).

Using our data set '**flight**', we can very easily plot two adjacent univariate scatterplots of the masses of pterosaurs and seabirds using base R's **stripchart** function:

```
stripchart(flight$mass_kg ~ flight$broad_group, ylim=c(-20,280),
  yaxs="i",method="jitter",jitter=0.2,vertical=TRUE,ylab =
  "Mass (kg)", pch=19,col="darkblue")
```

As with other plot types, we plot our y-variable against the x-variable (y~x); here for each sample (the x-variable in this case) we have only one quantitative measurement (mass). The bits of code specific to strip-charts are: **method**, where we here overwrite the default method **"overplot"** and specify instead **"jitter"** in order to separate the points out over a small x-axis range (useful where you have multiple identical or very close observations); **jitter**, where we specify the amount of jittering (or separation) applied; and **vertical=TRUE**, where we tell R to draw the strip-charts vertically rather than the default horizontal.

In the resulting plot (reproduced in Figure 6.9a) we can see that all of the masses are plotted individually, but some of the seabird values are still a bit overlapped despite our jittering. If you run the code yourself, you'll see that the jittering is random, so each time you run the **stripchart** function you'll get a slightly different arrangement of data points. Nevertheless, adding the jitter makes our data points clearer than they would be otherwise (try removing the **method** and **jitter** arguments to see the data points arranged in straight vertical lines). Whether the strip-charts in panel **a** or the boxplots in panel **b** are 'better' depends on what you want your audience to take away from the figure; certainly, the small sample sizes are more apparent in panel **a**. There is also the option to present both the strip-charts and the boxplots together—plotting the boxplot and then overlaying it with the strip-charts by including **add=TRUE** in the **stripchart** function.

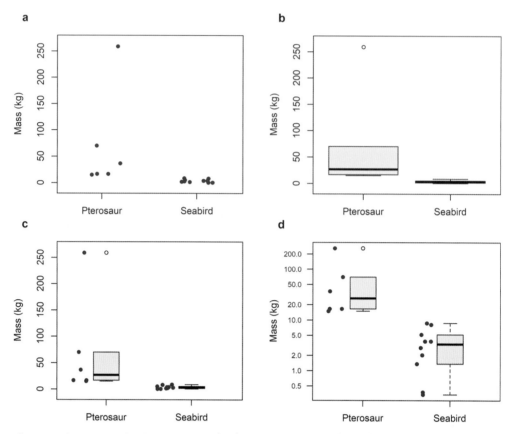

Figure 6.9 The masses of extinct pterosaurs (n=6) and extant seabirds (n=10) (Witton and Habib 2010) presented as: a) strip-charts, b) boxplots, c) boxplots overlaid by strip-charts, and d) boxplots overlaid by strip-charts on logarithmic y-axes.

We feel that adding strip-charts directly on top of the boxplots would look messy (particularly for the seabirds), so we present them here alongside one another (have a look at Figure 6.9c); to do this, we used the **boxwex** and **at** arguments when setting up our foundation boxplots (as previously seen in section 5.6.2 when preparing a grouped multiple-samples boxplot):

```
boxplot(flight$mass_kg ~flight$broad_group, ylim=c(-20,280),xlab = "",
    yaxs="i",ylab = "Mass (kg)",col="yellow",boxwex=0.3,at=1.3:2.3)
stripchart(flight$mass_kg ~ flight$broad_group, ylim=c(-20,280),
    yaxs="i",method="jitter",jitter=0.1,vertical=TRUE,ylab = "",
    pch=19,col="darkblue",add=TRUE)
```

What is apparent across panels **a–c** is that stretching the y-axis to include the outlier pterosaur mass of one of the *Quetzalcoatlus* makes it much harder to differentiate between the masses of our seabirds. This is fine if our point is that generally pterosaurs had greater mass, with some truly hefty individuals, but what if we were really interested in the individual points of both taxa? Our solution to this would be to use a logarithmic y-axis (see Bigger Picture 4.1 and section 8.5 for when and how to do this) for both our boxplots and strip-charts (like that in Figure 6.9d), though note that this could appear misleading if sufficient attention is not drawn to the unequal axis spacing, in the figure itself or its accompanying caption.

Whichever plot you prefer, pterosaurs certainly seem to have greater mass (and sometimes much greater mass) than our flying birds of today; but, given that their anatomy was completely different, we cannot suggest from this alone that pterosaurs were any less capable of flight than extant birds (Witton and Habib 2010).

Note: If you are interested in how we arranged the four panels in Figure 6.9, see the online resource 'R script for chapter 6' and section 7.2.2, where we discuss multi-panel plots in detail.

To conclude, strip-charts are most useful when presenting a small set of one-dimensional quantitative data. All of the individual data points are shown (as in multivariate scatterplots), and they can give a good sense of the spread of the data as well as any outliers. They can also be used alongside a summary plot of the data (such as a boxplot) where descriptive statistics as well as the individual data points are important to what you want to communicate.

 Key point

Strip-charts can be a great alternative or addition to a boxplot, if the sample size is small.

6.6   Non-zero lowest y-axis values on scatterplots

Throughout this chapter, we have specified the limits of our y-axes using the code `ylim`. For the puromycin data, for example, our y-axis ran from 0–250 DPM/min. However, not all scatterplots have to run from a baseline of 0 and, depending on your data, even if you do not specify `ylim` values yourself R will not always set 0 as the minimum y-axis value for scatterplots.

In section 3.7 we discussed occasions where starting the y-axis at non-zero values can be appropriate and, though this was written in the context of bar charts, much the same advice applies to scatterplots so we will not repeat ourselves here. However, non-zero values are more commonly seen in scatterplots, because they are used to present (sometimes large) quantitative data values rather than count data that would intuitively start at zero. Make sure you set axis limits effectively *but honestly*, so that your data is communicated well but not misleadingly (see Further Reading for links to pages discussing non-zero axes). In section 8.2, we provide an example of a non-zero y-axis scatterplot with an inset figure of the same data with a y-axis starting from 0–this is one way you can avoid misleading viewers.

 Key point

There can be good reasons for starting your y-axis at a non-zero value, depending on the aspects of the data you want to emphasize. However, you should consider whether there is any danger that this decision might lead to misleading the reader about other aspects of the data.

6.7   Stretch your understanding

If you want to practise some of the methods and stretch your understanding of the concepts discussed in this chapter, have a go at producing scatterplots for the following two data sets from the online resources. You can then compare your figures with our attempts and have a read of why we did what we did and what we thought of each other's efforts (available as online resources called 'Authors' attempts for chapter 6' and the accompanying 'R script for Authors' attempts chapter 6').

The first data set (bite_force.csv) is adapted from a table arranged by Hite et al. (2019). Hite et al. (2019) investigated the bite force of naked mole-rats (*Heterocephalus glaber*), a species with morphological and anatomical adaptations that predict strong bite forces; e.g. their skulls are relatively tall and wide, and they have impressive masticatory musculature. Naked mole-rats also have eusocial hierarchies, with dominant and subordinate castes, but the relationship between bite force and social caste had not previously been explored. The data we provide details the body masses (g) and maximum bite force (N) for both dominant and subordinate castes of naked mole-rat, alongside a range of other mammalian species. You have a choice of two questions to explore through visualizing this data:

1   How does the relationship between body mass and bite force in dominant and subordinate naked mole-rats compare to the relationship seen across the *other rodents* in the data set? To answer this, you'll want to think carefully about how to differentiate the two data points for the naked mole-rats and how to compare them to the general trend for the rodents alone.

2   How does the relationship between body mass and bite force in dominant and subordinate naked mole-rats compare to the relationship seen across *all of the mammalian groups* in the data set? This question is quite a bit trickier to plot, as the inclusion of such a range of species' masses and bite forces means you should consider exploring log values rather than the raw data for both variables. See Bigger Picture 4.1 and the online resource 'R script for chapter 4' to see how we plotted log-transformed data in panel **b** of BP 4.1 Figure A. Again, think carefully about what would be valuable to subset and how you can best compare the castes of naked mole-rat to the general trend across all the species. Check out our attempts at this in the online resources for some pointers.

The second example (core_stem.csv) is adapted from a figure by WISE Campaign (2019) reporting workforce statistics for the UK in 2019. WISE is a community interest company campaigning to 'increase the participation, contribution and success of women in science, technology, engineering and mathematics (STEM)'. The data we provide shows the total count (in millions) of workers overall, men, and women in core STEM roles from June 2009 to June 2019. We also provide the percentages that men and women make up in the core STEM workforce for each year—percentages like these can be treated as continuous data because the percentage can take any value from the continuum of zero to 100 per cent. Have a think about how best to demonstrate whether or not there has been an increase in women in STEM over the last 10 years.

For more scatterplot practice, R also has another couple of preloaded data sets you can look into. Try running the code below to learn more about 'ChickWeight' and 'CO2':

```
data(ChickWeight)
View(ChickWeight)
?ChickWeight
```

or:

```
data(CO2)
View(CO2)
?CO2
```

 Chapter Summary

- Scatterplots are used to explore whether there is a relationship between two variables, when those variables are continuous, or discrete quantitative variables with a logical ordering and at least seven different possible values, or a mix of the two.
- Using R, the basic code for scatterplots is: `plot(x, y)` but with additional code it is easy to modify points, control axes, add grid lines if desired, and plot multiple samples on the same figure.
- Lines of best fit, whether linear or curved, are commonly used with scatterplots to help visualize any correlations between variables or patterns in data.
- Time series are a common type of scatterplot to use where you want to plot change in one continuous variable over time—but remember to order the data by time if you want to connect the values together using `lines`.
- Univariate scatterplots (strip-charts) are most useful for one-dimensional quantitative data with a small sample size; they can also be used alongside summary plots.
- As with bar charts (discussed in chapter 3), there are advantages and disadvantages to starting scatterplot y-axes at a non-zero value; an inset can be a valuable compromise.

Online Resources

The following online resources are available for this chapter at www.oup.com/he/humphreys-obp1e:

- R script for chapter 6
- puromycin_R_data.csv
- land_and_ocean.csv
- LH_blood.csv
- birds_and_pterosaurs.csv
- bite_force.csv

- core_stem.csv
- Authors' attempts for chapter 6
- R script for Authors' attempts chapter 6

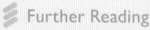 Further Reading

- 'Difference between LOESS and LOWESS' gives several answers to this question, with detailed information on the differences between the two: https://stats.stackexchange.com/questions/161069/ difference-between-loess-and-lowess
- For a tutorial showing how to use **lowess** to smooth lines and scatterplots in R, see 'lowess() R Smoothing Function | 2 Example Codes for Normalization by Lowess Regression': https://statisticsglobe.com/lowess-r-smoothing-function/
- 'Scatter Plot Smoothing' (https://stat.ethz.ch/R-manual/R-devel/library/ stats/html/lowess.html): we were determined to keep this book focussed on visualization, not statistics—but when it comes to plotting lines generated from the data in this chapter, this has been a tricky line to walk. There is a huge diversity of ways to derive a model that offers some summary of the 'shape' of the data, and lots of ways of picking a good one (and avoiding an inappropriate one) for your situation. 'Chapter 3: Model Fitting' from *Data Science with R* by Garrett Grolemund (https://garrettgman.github.io/ model-fitting/) and the book on *Fitting Models to Biological Data Using Linear and Non-Linear Regression* by Motulsky and Christopoulos (OUP, 2004) are good resources to help you explore this. Another useful resource is the page 'On Curve Fitting Using R' on Dave Tang's blog (https://davetang.org/ muse/2013/05/09/on-curve-fitting/)
- 'Truncating the Y-Axis: Threat or Menace?', an article by Correll et al. (2020), explores the effects on interpretation of not beginning the y-axis at zero: https://doi.org/10.1145/3313831.3376222
- 'Bar Charts Should Always Start at zero. But What About Line Charts?' looks at how to decide when a chart should start at zero, or when it would be more helpful to truncate the y-axis: http://www.chadskelton.com/2018/06/bar-charts-should-always-start-at- zero.html
- 'It's OK Not to Start your y-axis at Zero' argues for a number of situations in which truncating the y-axis is justifiable: https://qz.com/418083/its-ok-not-to-start-your-y-axis-at-zero/
- 'R Strip Chart' uses examples to demonstrate how to create strip-charts in R using the **stripchart** function: https://www.datamentor.io/r-programming/strip-chart/

CUSTOMIZING EVERYTHING USING R: DAY-TO-DAY

7

Learning objectives

By the end of this chapter you should be able to:

- Play with space in the R plotting area, by modifying graphical **parameters**.
- Customize design features of any plot to create visually appealing and informative figures.
- Add appropriate **legends** anywhere you want on a plot.
- Modify, or create entirely customized, axes in R graphs.

7.1 Introduction

This is the first of two chapters in this book dedicated to customization of graphs. In this chapter we will cover features and functions that crop up repeatedly throughout this book and that you will encounter often in your plotting of figures in R. There are some other things that we think are sometimes useful, but not on quite a day-to-day basis, and we save them for chapter 8.

More than previous chapters, we have designed this chapter to make dipping in quickly to find what you want as easy as possible. Thus, the sections will be more 'stand-alone', and this chapter might not have as strong a narrative thread running through it as previous ones. The section on each topic will provide you with some universal code you can apply yourself to any figure, and/or direct you to examples you might have encountered elsewhere in the book. The range of R code we describe is by no means exhaustive (the world of R code available online and in other books seems boundless!), but should give you a good grounding, and enable you to produce even higher quality figures exactly the way you want them.

 Key point

R is incredibly flexible. If there is anything about your figure that you would prefer to be different, then have a go at changing it to be the way you want. We bet you will succeed almost every time.

7.2   Graphical parameters with par

R has a vast range of customizable graphical parameters that enable you to control the way your figures are displayed. The function **par**–think 'parameters'–can be used to edit many features of your graphs, either for improved functionality or purely aesthetic purposes. Examples of all the different parameters controlled by **par** are available through one of the Further Reading links ('Graphical Parameters of R {graphics} Package'), but you could write a whole book on all these options! Here we will cover what we feel will be the most useful arguments for your day-to-day graphing.

7.2.1   Setting plot boundaries

R plots have two boundaries you can edit: the plot margins and the outer margin area. Generally, you might want to increase one or more of the margins to create some space for added detail close to the plot and related to the plot. You might want to increase the outer margins to add something more substantial not strongly related to the plot, like another plot entirely (see section 7.2.2 for details on multi-panel plots). You can control these boundary sizes by calling the **par** function before you run the commands for your plot and giving the corresponding arguments: **mar=c()** for margins, and **oma=c()** for outer margin area. For both arguments, you must list four values giving the desired space that you want to leave at the bottom, left, top, and right sides of the chart respectively. The values you give for your desired space should be based on the numbered 'lines' that R gives as size options for the space around these plotting areas (although equally spaced, these are somewhat arbitrary rather than being based on well-rounded measurements). To demonstrate this, Figure 7.1 shows the line placements that values 0–3 would correspond to for both the margin and outer margin area boundaries. As an example, the command **par(oma=c(0,4,0,0))** would prepare to draw a plot with a bigger than normal outer margin area on the left side of the plot only (in terms of R's understanding of the space, this outer margin area on the left side would extend to line 4).

The default margins (**mar**) R produces plots with are positioned with space values, in order, of 5.1, 4.1, 4.1, 2.1. This means there is more space at the base of the plot and the least space to the right of the plot. This makes sense as you often want space for details related to the x- and y-axes (like labels and values), and you might want a title above the plot too, but you don't ordinarily want to add much to the right of a figure. By default, there is no outer margin area on any side of the plot: **par(oma=c(0,0,0,0))**. Whenever you have customized the margins or outer margin area of a figure, a useful habit to get into is restoring the plot defaults afterwards if you plan to produce more figures in the same R session. There are lots of ways you could do this, but to us the simplest solution is simply to run commands redefining the boundaries:

```
par(mar=c(5.1, 4.1, 4.1, 2.1))
par(oma=c(0,0,0,0))
```

Note that, as an alternative to **mar** and **oma**, you could use **mai** and **omi** respectively, with the same code structure, if you would prefer to set the boundary areas by inches rather than lines. By default, **par(mai=c(1.02,0.82,0.82,0.42))** and **par(omi=c(0,0,0,0))**.

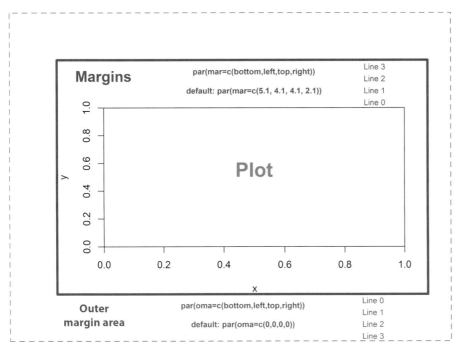

Figure 7.1 The code templates and defaults for margins (blue, solid line frame) and outer margin area (pink, dashed line frame) when plotting. The line values 0–3 indicate the placements of boundaries relative to the plotting area when those values are given. Figure inspired by https://www.r-graph-gallery.com/74-margin-and-oma-cheatsheet.html.

7.2.2 Multi-panel plots

It is very simple to arrange multiple plots in the same plotting space in R using the **par(mfrow)** parameter. Within **mfrow**, you simply need to list two values: the first is the number of rows you want your plots arranged in, the second is the number of columns you want your plots arranged in. For example, in section 4.3.2.2 we used the line of code **par(mfrow=c(1,3))** ahead of producing a figure containing three separate histogram plots, arranged horizontally as a single row with three columns. Similarly, in section 5.4.1 we used **par(mfrow=c(3,1))** to arrange three histograms vertically as a single column with three rows. The **mfrow** code should be followed by the code for each of the plots you want to include, and they will appear in order as panels positioned in the designated row–column structure. Importantly, although plots are being added to the same overall figure in the plotting area, the commands for each of the plots should not be separated by **par(new=TRUE)** as we do when adding grid lines or multiple samples to plots (see section 7.2.3 and cross-references therein), as we need R to plot them sequentially in their separate designated positions rather than adding them all on top of each other. To return the **mfrow** parameter to its default once you have finished producing a multi-plot, you need to specify **par(mfrow=c(1,1))**—that is, return to plotting a single figure (one row, one column) in the plotting area.

The **mfrow** parameter alone is usually enough to give you a clear multi-panel plot when you only want to include a few different plots. However, when you want to present more than three plots in the same figure, or when you want the figures you create using mfrow to be more compact, you can play with the plot boundaries (as we did in section 7.2.1) to make sure that all axes and all the information within each plot are clear. For several of the figures that we produced for chapter 6, we edited the margins of plots within multi-panel plots using mar so that we could include lots of plots in the same figure (see section 7.2.1 for more details on mar). For example, the online resource 'R script for chapter 6' details the code we used to produce BP 6.1 Figure A, including:

```
par(mfrow=c(2,3))
par(mar=c(4.1, 2.1, 2.1, 2.1))
```

BP 6.1 Figure A has a multi-panel plot of six possible scatterplot relationships arranged in two rows and three columns. Remember from section 7.2.1 that the default margin spaces are mar=c(5.1, 4.1, 4.1, 2.1).

In this figure we reduced the size of the margins at the bottom, the left, and the top of all the plots to help them sit together more compactly; this worked well for these plots as they did not have axis ticks or y-axis labels, only an mtext label below the x-axis that we needed to keep space for (see section 7.3.5 for more use of mtext). If you look at the 'R script for chapter 6', you will see that we also returned both of these parameters to their defaults after running the code for the six plots:

```
par(mar=c(5.1, 4.1, 4.1, 2.1))
par(mfrow=c(1,1))
```

Similarly, in section 6.5 we present a 2 by 2 multi-panel plot showing boxplots and strip-charts. The code we provide in the online resource 'R script for chapter 6' shows that we prepared our plotting area and boundaries with:

```
par(mfrow=c(2,2))
par(mar=c(2.1,4.1,2.1,2.1))
```

Here, we made our left-side margin bigger than the other margins for each plot; this is because the y-axis ticks and labels needed more space to be clearly presented. Again, we returned mfrow and mar to their defaults after producing the figure. See also the use of mfrow, mar, and oma in the code for Figure 7.6 at the end of this chapter, in section 7.5.2.

Sometimes when preparing multi-panel plots, the variables measured along your x- or y-axes might be the same across all the plots. Or you might have plots where the same colours are used across plots to provide the same additional information. In cases such as these, you might want to produce a multi-panel plot with shared axis labels or a shared legend. To do this, you can use the parameter oma to create some outer margin area for your plots (discussed more in section 7.2.1). In the online resource 'R script for chapter 6' you can see how we used oma as well as mfrow and mar to produce SA 6.3 Figure A:

```
par(mfrow=c(3,1))
par(mar=c(2.5,4,2.5,1))
par(oma=c(1.75,0.5,1.25,0))
```

The resulting figure has three plots positioned vertically in a single column. All plot margins are reduced from their defaults, but the left-hand side of each plot is kept a little wider than the others for the y-axis tick labels. However, we specify some outer margin spacing: i) at the bottom of the plots for the shared x-axis label to sit below the bottom plot, ii) at the left-hand side for the shared y-axis label to run along all the plots, and iii) at the top so the shared legend has space to sit above the top plot. We applied the same boundary parameters to all of the plots in our multi-panel plot in order to easily keep all the plots the same size, but you could also set either of the boundary parameters differently ahead of each plot if you wanted.

For SA 6.3 Figure A, you can see in the 'R script for chapter 6' that the shared legend is added after the code for the top plot, to easily position it outside of the plotting area at the top of the entire figure (see section 7.4 for more legend customization). If we had wanted to add a shared title to this multi-panel plot instead of a legend, the principles of how to do this would be the same as above, except we would have used the `main` argument or `mtext` function instead of the `legend` function; as a reminder, though, most scientific figures should be accompanied by a detailed caption instead of a title (and we provide advice on writing these in section 1.2 and Scientific Approach 1.2).

The shared y-axis is then added after the code for the middle plot using `mtext` to position it at the middle of the overall figure. The shared x-axis is added after the code for the bottom plot using `mtext` to position it centrally at the bottom of the overall figure. What feature you want to share between plots will affect the spacing you assign to different sides of plots and when in your code for the overall figure you position the code for that feature (for another example, see section 7.5.2 for the code for Figure 7.6). Sometimes figuring out the best line spacing for margins and placement of `mtext` can take a bit of trial and error, but with a little persistence it won't take you long to find values that work for your figure. The use of `mtext` to position text in figure margins or the outer margins of the device region is explained further in section 7.3.5.

Using `mfrow` and adjusting `mar` and `oma` as needed should meet the vast majority of your graphing needs in R. However, if you are interested in creating customized multi-panel plots of varying sizes, we recommend you investigate the links on the `layout` function provided in the Further Reading. Alternatively, the parameter **fig** can also be used for fine-scale control of plot locations, as we will discuss in section 7.2.3.

7.2.3 Completely controlling the space

As mentioned above, the graphical parameter **fig** can be used to precisely control the location of multiple figures within a plot. The entire plotting area that R works within has both its x- and y-coordinates running between 0 and 1. Therefore, the default **fig** parameter, wherein a plot fills the entire available plotting area, is given as:

```
par(fig=c(0,1,0,1))
```

The coordinate ranges for the x- and y-axes in **fig** are given in the order: x1, x2, y1, y2. Therefore, to position multiple plots in the plotting area at the same time we can figure out which coordinates within the ranges of 0 and 1 we can use to position plots wherever we want them. For example, to arrange two hypothetical plots side by side horizontally, instead of using `par(mfrow=c(1,2))` we could instead use the code:

```
par(fig=c(0,0.5,0,1))
plot1code
par(fig=c(0.5,1,0,1), new=TRUE)
plot2code
```

The first plot would fill the left-hand side of the plotting area by running between 0 and 0.5 on the x-axis, while the second plot would fill the right-hand side of the plotting area by running between 0.5 and 1 on the x-axis. Both plots would fill the full height of the plotting area, as they both reach from 0 to 1 on the y-axis. We could also produce exactly the same arrangement as `par(mfrow=c(2,1))` if we run the code:

```
par(fig=c(0,1,0.5,1))
plot1code
par(fig=c(0,1,0,0.5), new=TRUE)
plot2code
```

With the above code, the first plot would fill the top half of the plotting area by running between 0.5 and 1 on the y-axis, while the second plot would fill the bottom half of the plotting area by running between 0 and 0.5 on the y-axis. For any plot, you could assign any coordinates between 0 and 1 for both the x- and y-axes. Here, both plots would fill the full width of the plotting area, as they both stretch from 0 to 1 on the x-axis. Such control of the plotting space means that the arrangement of plots can be much more flexible than with `mfrow`, and plots are allowed to vary in size.

If you are interested in arranging plots of various sizes, see section 8.2, where we use the parameter **fig** to produce a large bar chart with a smaller, inset bar chart in the top right-hand corner of the plotting area. Also, check out section 7.5.2 at the end of this chapter for the code for Figure 7.6, where we use `mfrow` to first position two upper plots and then **fig** to position a single plot more centrally below, and subsequently use **fig** again to create a dummy plot on which we prepare shared axes (see also section 7.2.2 for more on shared axes). Importantly, when using **fig** to arrange multiple plots within the same plotting area, you must remember to include the code `new=TRUE` to every plot subsequent to the initial plot.

We use `par(new=TRUE)` on its own in other chapters, both when adding grid lines to figures (see examples in chapter 3, and sections 4.3.5, 6.3, 6.4, and 7.3.4) and when overlaying existing plots with additional plots of different samples (see examples in sections 5.4.2, 5.6.2, 6.3.2, and 6.4). As in those cases, the `new=TRUE` argument must be added to the **par** function when using **fig** to tell R that we want to add the subsequent plots to the same plotting area as the original plot, rather than overwrite the initial plot (see section 7.5.2 for our use of `par(new=TRUE)` in the code for Figure 7.6).

Remember also, when you have finished producing your plot(s) with customized layout, to restore the default **fig** parameter so that subsequent plots will fill the available plotting area again:

```
par(fig=c(0,1,0,1))
```

7.2.4 Establish graphing conditions

As with the parameters **mar**, **oma** (discussed in section 7.2.1), **mfrow** (discussed in section 7.2.2), and **fig** (discussed in section 7.2.3), other components of graphs in R can be customized for the duration of plotting by specifying different values in the **par** function. For example, many of the design features discussed below in section 7.3, can be established with a **par** command and will continue to apply to any figures you create until default values for those features are restored. For example, in section 3.6 we defined the line widths of an overly customized bar chart by running `par(lwd=3)` ahead of running the code for the bar chart itself. This drew the bar chart with all the bars' line widths set to thicker than usual (the default line width is 1). If we were to continue producing figures in the same session of R, the line widths of everything further we plotted would continue to be set to 3, until we redefined the line width for the session as the default value with `par(lwd=1)`.

The only way we could draw plots using the default line width without restoring the default using **par** would be to specify so in the code for specific plots—unlike the parameters discussed previously, the arguments for design features (such as those covered in section 7.3) are often customizable within the code for plots themselves rather than being defined with a separate **par** command. It is usually simpler to include customized arguments for design features within the code for plots or plot features in this way, so that you do not have to remember to return to the default parameter values.

 Key point

The R function **par** is a powerful and flexible tool for getting your figure exactly the way you want it—this section only covers what we feel are some common uses.

7.3   Design features

7.3.1 Colour

An obvious feature you might want to customize, R has an amazing range of colours you can use when graphing. In the online resources for chapter 1, we provide a handy one-page document called 'Colour Guide' that lists some of the colours available in R (including all the colours used in this book). See also Further Reading in chapter 1 for links to guides on Brewer palettes, if you are interested in using a stylish package to select appropriate colours. If you want to choose from all of the possible colours available in R, you can find a full list at: http://www.stat.columbia.edu/~tzheng/files/Rcolor.pdf

The basic code to select a given colour in R is simply: `col=`. You can then name any single colour you want by simply using R's quotation marks (e.g. `col="blue"`) or a list of whichever colours you want in the order you want them, applied using R's typical list format, e.g. `col=c("blue", "green", "mistyrose")`. Colours can be specified via their index number, hexadecimal code, or name. For example, the colour named `"white"` is also indexed as `1` or can be referred to with its hexadecimal code `"#FFFFFF"`.

But throughout this book (and in our own graphing work) we prefer to stick to colour names—in particular, names make it easier to check that your colours are correctly positioned when you are using multiple colours as a sequential or diverging palette (different colour palettes are discussed in section 1.6 and Scientific Approach 1.3). Variations on the code to specify colour exist for some different graphing features, though. For example, for different text components in your plots you can specify:

> col.axis to customize the colour of axis annotations (see section 7.5.2 for how this is used in the code for Figure 7.6)

> col.lab to customize the colour of labels associated with the x- and y-axes

> col.main to customize the colour of font in a figure title (as we demonstrate in sections 3.2 and 3.6)

> col.sub to customize the colour of font in a subtitle.

See Figure 7.2 and its associated code (included in sections 7.3.2 and 7.3.3) to see some of these variations in use. Other graphing components, such as legends with colour blocks (covered more in section 7.4) and point shapes with backgrounds (covered more in section 7.3.3), also have variations on how to specify colour (**fill** and **bg**, respectively) but these are covered in more detail at relevant points throughout the book.

Rather than repeat ourselves here, see section 1.6 for our advice on when to use colour in figures and how to effectively select appropriate colours.

7.3.2 Font size and style

Although R's default font sizes for different graph components are usually appropriate, they can easily be edited if you need to; for example, you might have qualitative categorical variables with many or long names that all need to fit on a single figure, or you might want some large **mtext** to act as a shared axis label alongside many plots (see section 7.2.2 for details on multi-panel plots). To edit font size, **cex** is the go-to argument to include in your code. By default, font size is set to **cex=1**, so if you wanted your font to be 50 per cent smaller you could specify **cex=0.5**, or if you wanted it 50 per cent larger you could specify **cex=1.5**. As with colour, there are variations to **cex** based on the component of the figure you want to change the size of, including:

> cex.axis to customize the size of axis annotations (see section 7.5.2 for use of this in the code for Figure 7.6)

> cex.lab to customize the size of labels associated with the x- and y-axes

> cex.main to customize the size of font in a figure title

> cex.sub to customize the size of font in a subtitle

> cex.names to customize size of the names of bars in bar charts.

R also has several options for changing the font 'face', where you might not want to change the size of font but you still want to emphasize it or alter its look in some way. The options for font face in R are really simple to call. Simply include **font=** in your command and specify either 1, 2, 3, or 4 where:

> 1 = normal (the default)

> 2 = **bold**

> 3 = *italic*

> 4 = ***bold and italic***

Again, specialized versions of the font face code include:

> font.axis to customize the font face of axis annotations (again, see section 7.5.2 for use of this in the code for Figure 7.6)

`font.lab` to customize the font face of labels associated with the x- and y-axes

`font.main` to customize the font face of a figure title

`font.sub` to customize the font face of a subtitle.

A final font customization option in R is to change the font 'family'. The default font family in R is 'sans', but base R also contains the families 'serif' and 'mono'. To change the default font family a plot or some feature of a plot is drawn in, run the command:

`par(family=" ")`

Remember to specify '**serif**' or '**mono**' in the quotation marks. You can then run whatever code where you want that font family to apply, before restoring the default: `par(family="sans")`. Font family is an example of a feature where **par** should be used to establish the customized graphing conditions (rather than it being an argument you can easily include in the code for a plot), as discussed in section 7.2.4.

Check out the code below and the associated Figure 7.2 to see what the three default fonts look like, and notice some font size and style customization code in action:

```
par(family = "serif")
plot(1, type="n", xlab="serif family, face=4 x-axis",
  ylab="serif family, face=4 y-axis",
  cex.lab=1.2,col.lab="purple4",font.lab=4,col.axis="red",
  font.axis=2,
  main="Colour, font and point customisation",
  cex.main=1.5,col.main="forestgreen",
  font.main=3,xlim=c(0, 1), ylim=c(0, 1),xaxs="i",yaxs="i")
par(family = "sans")
text(0.2,0.9,labels="Font families:")
text(0.2,0.75,labels="'This is sans'",cex=1.5)
par(family = "serif")
text(0.2,0.6,labels="'This is serif'",cex=1.5)
par(family = "mono")
text(0.2,0.45,labels="'This is mono'",cex=1.5)
par(family = "sans")
```

It is also possible to specify other font families, but the means by which this is achieved varies between devices and output formats so, for this reason, we cover only the families available in base R; for more information on the ways you might be able to use other font families see the Further Reading.

7.3.3 Point shapes

In chapter 6, we provide SA 6.1 Figure A, detailing all the point shapes available in R, because scatterplots are primarily where you will need to add points to figures (usually just via the **plot** function). However, to demonstrate some point types (specified using **pch**) and the use of the **points** function (which allows you to add points to any specified coordinates), we provide some code to add onto that from section 7.3.2 to create Figure 7.2 in its entirety:

```
text(0.65,0.85,labels="pch= 0, 1, 2",cex=1)
points(0.55,0.75,pch=0,col="red")
```

```
points(0.65,0.75,pch=1,col="purple4")
points(0.75,0.75,pch=2,col="forestgreen")
text(0.9,0.75,labels="outline",cex=0.8,font=2)
text(0.65,0.55,labels="pch= 15, 16, 17",cex=1)
points(0.55,0.45,pch=15,col="red")
points(0.65,0.45,pch=16,col="purple4")
points(0.75,0.45,pch=17,col="forestgreen")
text(0.9,0.45,labels="filled",cex=0.8,font=2)
text(0.65,0.25,labels="pch= 22, 21, 24",cex=1)
points(0.55,0.15,pch=22,col="red",bg="yellow")
points(0.65,0.15,pch=21,col="purple4",bg="orange")
points(0.75,0.15,pch=24,col="forestgreen",bg="skyblue")
text(0.9,0.15,labels="outline and\nbackground (bg)",
  cex=0.75,font=2)
```

Note: the use of \n tells R to start a new line at this point in the text—you will see that the \n itself is not printed on the resulting figure. See sections 3.5.1 and 8.7 for more uses of \n for effective text placement.

The points that we show here display three variations on the basic shapes square, circle, and triangle. The first three versions **pch** offers (0, 1, and 2 respectively) draw these shapes with an outline alone. The second three versions **pch** offers (15, 16, and 17 respectively) draw these shapes filled in with the same colour as their outlines. And the final three versions of these shapes **pch** offers (22, 21, and 24 respectively) draw these shapes with the outline colour specified by **col** and a different interior fill (or background) colour specified by **bg**. As a reminder, all of the point types available with **pch** are shown in SA 6.1 Figure A.

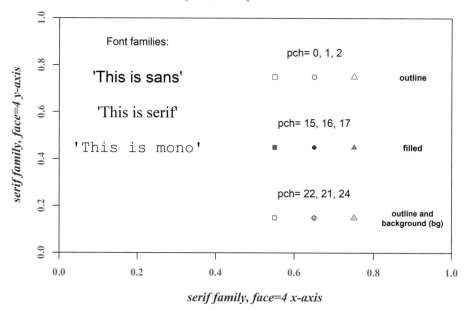

Colour, font, and point customization

Figure 7.2 Some of the many colour, font, and point customization options available in R. This figure was produced by code provided in both sections 7.3.2 and 7.3.3. Colours, font sizes, and styles vary across the code. The base plot title, axes, and labels are drawn with the font family 'serif', and all three base R font families are shown on the left-hand side of the plot. Variations of square, circle, and triangle point shapes are shown on the right-hand side of the plot.

7.3.4 Lines and grid lines

The lines used to draw features of some plots, or additional lines you may choose to add to your graphs (e.g. grid lines, a line of best fit, to join data points, or to indicate some kind of reference threshold), can also be customized in R. Colour can be specified with `col` in the same way we have seen in section 7.3.1.

Line types are specified with `lty`, with the following number producing a line of a specific style (as we show in Figure 7.3). You could also specify line type using text if you prefer, e.g.:

```
lty=c("blank", "solid", "dashed", "dotted", "dotdash",
"longdash", "twodash")
```

By default, line type is 1 (`"solid"`). Line width is specified using `lwd` and a number >0. By default, `lwd=1`, but see Figure 7.3 for examples of line widths 1–8. See the Further Reading for even more fine-tuning of line types, including line end and line join styles.

It is simple, then, to customize lines, but how can lines be added to graphs in R? Straight lines that span the plotting area can be added to graphs using the `abline` function. If you want to add a vertical line running up from a specific x-axis coordinate (as we demonstrate in section 6.3.3—see also Figure 7.4 for another example), the structure of the function is simply:

```
abline(v = x)
```

If you wanted to add a horizontal line running along from a specific y-axis coordinate (as we do in Figure 7.4), the structure would be:

```
abline(h = y)
```

Of course, the x- or y-value given in these cases could also be a value that you get R to calculate, rather than a value you know precisely yourself, e.g.:

```
abline(v=mean(mydata$xvariable))
```

Sometimes, rather than a 90° line, you might want to add a sloped line to your data, particularly if you are showing an overall trend (most likely on a scatter-plot). In cases such as this, the structure of the `abline` function would be:

```
abline(a, b)
```

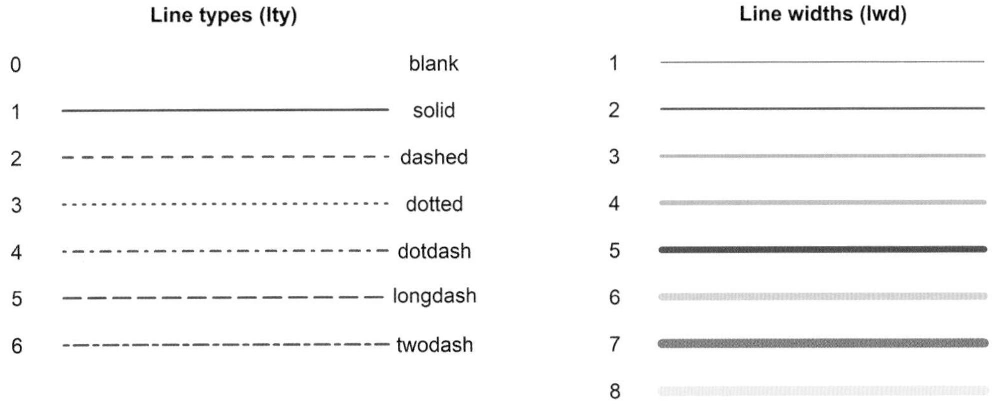

Line types (lty)

0	blank
1	solid
2	dashed
3	dotted
4	dotdash
5	longdash
6	twodash

Line widths (lwd)

1	
2	
3	
4	
5	
6	
7	
8	

Figure 7.3 The values and names corresponding to different line types in R (`lty`) and the line widths (`lwd`) 1–8.

where **a** and **b** are single values that specify, respectively, the intercept and slope of the desired line (see an example in Figure 7.4, where a=0.3 and b=10.5).

More commonly, you may use **abline** to easily add a regression line (or linear trend line) to your data. We do so in sections 6.3.1 and 6.3.2, but essentially **abline** is used in conjunction with **lm** (the linear model function) to achieve this:

```
abline(lm(y~x))
```

In sections 6.3.3 and 6.4 we also explain how to use the **lines** function to add curved trend lines to your plots, or to connect points in a time series. **lines** takes any coordinates you provide (it accepts different forms of data, e.g. lists or data sets) and joins the corresponding points with line segments (see also section 7.5.2 for use of **lines** in the code for Figure 7.6). In the code, you must provide the x- and y-coordinates, and you can specify a 'type' of line if desired:

```
lines(x, y, type = "l")
```

By default the **type** is **"l"** for lines. But, as shown in Figure 7.4, you could also specify:

"p" to draw points alone

"b" to draw both points and lines

"c" to draw empty points joined by lines

"o" to draw overplotted points and lines

"s" to draw stair-like steps.

A final option you can use to draw lines in R, the **segments** function, allows you to draw a straight line segment between two points anywhere on a graph without it needing to span the entire plotting area (as **abline** does); see Figure 7.4. The **segments** function is simply structured:

```
segments(x0, y0, x1, y1)
```

where **x0** and **y0** are the coordinates where you want the line to start from and **x1** and **y1** are the coordinates where you want the line to end. In Scientific Approach 6.2 we explained how **segments** can be used to draw a restricted line of best fit onto a scatterplot using values extracted from a linear model.

For any type of line you draw with **abline**, **lines**, or **segments**, the line type, width, and colour can be customized by including the arguments outlined above in their line of code (see the online resource 'R script for chapter 7' for the customizations we apply to Figure 7.4).

A common reason you might want to add lines to a graph is to provide grid lines behind the plotted data to help viewers read precise values from the graph. Where figures are used to present overall trends in the data, or precise values are given in accompanying text, grid lines are not usually necessary and can contribute to 'chart junk' that instead clutters up the visual appearance (see section 1.7 for more discussion on 'chart junk'). However if, for example, you had data plotted on a scatterplot that showed the relationship between an x- and a y-variable, you might also want viewers to be able to predict y-values based on x-values that were not tested during data collection—this would be easier for them to determine if they had reference lines at standard intervals for each axis.

Rather than repeat ourselves here, we demonstrate how to use **abline** and the customization arguments shown above to add grid lines to bar charts in chapter 3, histograms in chapter 4, and scatterplots in chapter 6. The same

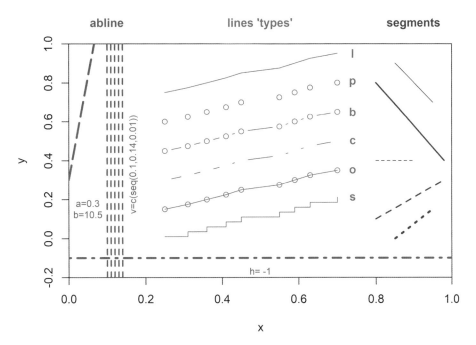

Figure 7.4 Variations of the different ways to draw lines in R: `abline`, `lines`, and `segments`.

principle could be used to add grid lines to boxplots (chapters 4 and 5) or any other kind of graph that you might plot that has a quantitative x- and/or y-axis.
The basic steps in any instance are:

1   Create a blank/dummy plot with no data in the plotting area, but the same x- and y-axis limits as will be defined in the final full graph.

2   Establish grid lines along the axis/axes as required using `abline`—we tend to use a '`lightgray`' colour for minor grid lines and a '`darkgray`' colour for major grid lines (keeping the default `lty=1` and `lwd=1` for both).

3   Run the code `par(new=TRUE)` to tell R to add the following code to the existing plot rather than to overwrite the existing plot.

4   Run the code for your full graph so that it is produced on top of the prepared grid lines.

7.3.5 Text

Adding text anywhere you want to a graph is very simple in R, and in fact we have already included code that adds text to graphs earlier in this chapter (see sections 7.3.2 and 7.3.3). The basic `text` function is structured as `text(x, y, labels=" ")` where `x` and `y` are the coordinates, or lists of coordinates, where text is to be placed, and `labels` is followed either by a single piece of text or value, or a list of text/values of the same length as the lists of x- and y-coordinates (this can just be the pre-existing name assigned to a list defined earlier in your code).

The examples of `text` in the code earlier in the chapter (see sections 7.3.2 and 7.3.3) each position only a single text label onto Figure 7.2. But in section 5.3.2 we provided a list of x-coordinates and a pre-existing list of sample sizes as the `labels` in order to add the corresponding sample size below each box of a boxplot.

Sometimes, we do not even need to specify the values or coordinates in the code ourselves if they are available by some other means. For example, in section 6.3.3 we positioned the values from an existing variable 'rate' on a scatterplot at the corresponding coordinates extracted from the relationship between the existing variables of rate and concentration ('conc'): `text(rate~conc, labels=rate)`.

As with other functions in R, calculations of values to be printed can also be included within the `text` function. As an example, `text` is used again in section 6.3.3 to position the mean of two values on a scatterplot (Figure 6.6) by including the following under the `labels` argument: `paste("mean =",mean(c(191,201)))`. Used with `text`, this command 'pastes' the text 'mean=' followed by the mean value '196' that R calculates within the command onto the graph. Similarly, the `round` function could be included within the `text` function if any values to be added to a graph need to be rounded to a sensible number of decimal places. The structure of `round` is `round(x, digits=)`, where `x` is the numeric value or list of values (or the code for a calculation, such as `mean`, contained in parentheses) to be rounded off, and `digits` is the number of digits to which any values have to be rounded off (see the Further Reading for more explanation).

We used `round` in section 2.2.1.2 to calculate the relative percentages of segments in a pie chart to the nearest one decimal place. Again, `round` could be used in conjunction with the `paste` function if there is to be any sort of text accompanying the value (or values). See the relevant link in the Further Reading, and also section 8.6 and Scientific Approach 8.1, for further use of the `paste` function (and the use of `expression`) in commands involving `text`, including the use of mathematical symbols, superscript, and subscript.

Within the code for `text`, the size, colour, and font can all be customized with the arguments outlined earlier in this chapter, as you will find in examples throughout this book. Text could also be positioned outside the main graph by specifying coordinates that would sit outside the x- and y-axis ranges and by including the argument `xpd=TRUE` in the `text` function (this simply tells R to allow `text` to function outside the plotting region). However, if you want to place text outside the plotting area at consistent locations across different graphs with varying axes, `mtext` may be a better option.

Although `mtext` can be used to position text within the plotting area (by defining negative value `line` positions) it is most commonly used to write text in one of the four margins or outer margins of a figure (as discussed in section 7.2.1). The basic structure of the `mtext` function is:

```
mtext("text", side =, line =)
```

where `text` is whatever text you want to position contained within R's quotation marks " ", `side` is which side of the plot the text should be positioned (1=bottom, 2=left, 3=top, 4=right), and `line` is on which margin line (remember `par(mar)` from section 7.2.1) you want the text to be positioned (see Figure 7.1 to see this in action).

Because the `mtext` function bases its `line` argument on the plot margin alone (and not the outer margin area, `par(oma)`), if you want the text to be positioned beyond the margin area of your plot you will also need to include either the code `outer=FALSE` (which tells R to use outer margins if available) or xpd=TRUE (which, as for **text**, allows R to plot outside the limits of the plot and its margins). Optional additional arguments for **mtext** include:

> **at** = can be used to specify the x- (if on sides 1 or 3) or y-coordinates (if on sides 2 or 4) the text should be positioned alongside, although this is best excluded if you want the **mtext** position to be consistent across multiple graphs that vary in their axis ranges (unless you add a blank dummy plot with consistent axes across all plots before adding **mtext**).

> **las** = specifies whether text should be positioned parallel to the axis on its side of the plot (0, default), horizontal (1), perpendicular to the axis on its side of the plot (2), or vertical (3).

> **adj** = adjusts the position in the reading direction of the text, where 0 means left or bottom alignment, and 1 means right or top alignment.

> **padj** = adjusts the position perpendicular to the reading direction of the text, where 0 means right or top alignment and 1 means left or bottom alignment.

As touched on in section 7.2.2, **mtext** is the function to use when you are creating multi-panel plots that require shared axis labels or titles, due to its easy positioning along the sides of any plot or plots (see the online resource 'R script for chapter 6' for our use of **mtext** when producing SA 6.3 Figure A).

It is also useful for positioning letters or numbers to label each panel within a multi-panel plot, as—through **mtext**'s use of plot margin lines—these labels can be positioned in exactly the same location relative to each of their respective plots. We use **mtext** for exactly this purpose in section 7.5 at the end of this chapter—you can see in the code provided that the **side**, **line**, **adj**, **cex**, and **font** arguments have the same value for positioning letter labels at each of panels a, b, and c when producing Figure 7.6.

However, as mentioned earlier, although **mtext** is useful for adding text outside the plotting region, it can also be used to position text within the plotting area; to do this, you simply have to specify negative value **line** positions. Again, this could allow for consistency in text placement across figures with varying axes, but see the code for SA 6.3 Figure A in the online resource 'R script for chapter 6' for an example of in-figure **mtext** placement with consistent axes across panels.

 Key point

You have so much freedom in changing the look of your figure in R—it takes just a little extra effort to make your figure clearer, easier to assimilate, and more eye-catching.

7.4   Legend design and placement

Whenever a figure includes data on an important variable that is not shown on an axis, a legend is really helpful to allow viewers to interpret this figure. That is, when you use different colours, points, and/or line types, you will often want to include a key to that feature that explains the meaning of the differences.

Another option might be to provide this explanation in the figure caption, but a legend is often a better choice unless the explanation is very simple.

In this book we use legends when graphing qualitative, categorical data as either pie charts (see chapter 2) or bar charts (see chapter 3), but also when plotting multiple samples of quantitative data onto the same graph, whether the data is displayed as a histogram (see section 5.4.2), boxplot (see section 5.6.2), or scatterplot (see section 6.3.2).

The basic code structure when using the **legend** function is:

```
legend(location, legend=, fill= or col=)
```

The location of the legend can specified using any one of the keywords: `"bottomright"`, `"bottom"`, `"bottomleft"`, `"left"`, `"topleft"`, `"top"`, `"topright"`, `"right"`, and `"center"`. As an optional argument, you can also provide an **inset** value to specify how far the legend should be pulled in from the plot margins. The effects of using some of these keywords (and optional inset) are shown in Figure 7.5.

Alternatively, you can position a legend more precisely by specifying coordinates (in the order **x, y**) as the location rather than using one of the standard keywords. Legends can even be positioned outside of the plotting area, by including the argument **xpd=TRUE**, though in this case usually you will want to create some extra margin space (**mar**) or outer margin area (**oma**) for it in advance of plotting your figure (see details of **mar** and **oma** in section 7.2.1).

The other main arguments in **legend** relate to the variables you are presenting via a design feature (e.g. colour, point shape, line type). Using the argument **legend** (confusingly within the function **legend**!) you should create a list of text that explains the differences in the design feature shown/the levels of the variable. If you need to produce square blocks of colour within the legend, as you would for a pie chart and might for a bar chart, boxplot, or histogram, then **fill** is the argument you use to specify the colours needed (see sections and chapters named above). If you have used different colours in a scatterplot you would use **col** to identify their meaning. With scatterplots it is also possible to differentiate point shapes, line types, and/or line widths; these should be

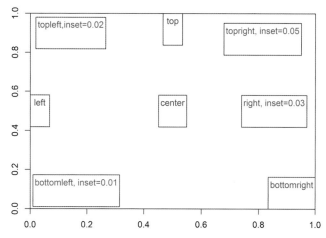

Figure 7.5 Standard legend locations in R, with blue text indicating default locations and pink text indicating default locations that have had their positions slightly altered with the argument **inset**. Figure inspired by http://www.sthda.com/english/wiki/add-legends-to-plots-in-r-software-the-easiest-way

explained in the legend using `pch`, `lty`, and `lwd` respectively. Remember that for all the components of your legend, you should keep the order in which you list levels of the variable and any variations of a design feature the same, so that the correct combination of level label and design feature(s) is produced in the final legend. For an example, see section 6.3.2, where we add a legend to a scatterplot that explains the use of two different point shapes and two different colours for two label levels using the commands:

```
pch = c(15,19), col = c("orange", "blue"), legend=c("Treated",
"Untreated")
```

Here the design components for the 'Treated' data are always the first listed (point shape 15 and colour orange), while the design components for the 'Untreated' data are always the second listed (point shape 19 and colour blue).

Additional arguments you might commonly want to include in your code for legends include:

> `inset` = mentioned above, the value given here specifies how far the legend should be pulled in from the plot margins from its specified location
>
> `bty` = this means 'box type', and by specifying `"n"` you can remove the outline of the box from your legend
>
> `title` = this can be used to add the variable whose levels are differentiated by colour, point shapes, line type, etc. as a heading within the legend
>
> `bg` = this can be used to specify a background colour for a legend box— we specify `bg="white"` when adding a legend to a graph that has grid lines (see section 7.3.4 for details on grid lines), but this is not usually necessary.

To demonstrate some of the customizable features of legends, we produced one for each panel of Figure 7.6 (though they were not necessary in panels **a** and **b**). You should find it is pretty easy to match each of the following lines of code to their corresponding panel in Figure 7.6:

```
legend("topleft",pch=16,cex=1.5,col="red",legend="sample 1")
legend("topleft",bty="n",inset=0.075,cex=1.5,lty=1,
  col="red",legend="sample 1")
legend("topleft",bty="n",cex=1.5,lty=c(1,2),col=c("red","royalblue"),
  pch=c(16,17),legend=c("1","2"),title="Sample:")
```

 Key point

In all but the simplest of figures, you will need a legend to explain some of the informational content of the figure. As ever in R, you can make sure you produce the legend in exactly the way you want.

7.5   Customizing axes

7.5.1 Modifying automatic axes

The axes R automatically produces for graphs are based on the data plotted and usually look pretty good. In Figure 7.6a, we produced a plot of some simple data that we generated relying on R's automatic axes and they managed to fit the data comfortably.

However, the axis limits do not always cover the range of x- or y-coordinates you would like them to. To specify the limits of the x-axis you simply need to include xlim=c(minx,maxx) when plotting, where minx and maxx are the minimum and maximum ranges you want your x-axis to run between. To specify the limits of the y-axis, the required argument is ylim, following the same structure: ylim=c(miny,maxy).

Even when these axis limits are applied, the first and last tick marks of the axes do not always neatly reach the end limits of the axes. If this is something you would like to neaten, this can usually be achieved by including the code xaxs="i" and/or yaxs="i"; this is what we did when producing Figure 7.6b. However, sometimes this still does not manage to force axes exactly where you want them and, in cases such as this, you might choose to draw new axes altogether (as we discuss in section 7.5.2).

An additional customization option available with the automatic axes in R is las, which we explained in section 7.3.5 in the context of mtext. las can be included in the main code for a graph and specifies how axis tick mark labels should be oriented with respect to the axes. You could also use log when plotting to produce logarithmic x- and/or y-axes, but this is covered in detail in section 8.5.

As a final easy piece of code to customize R's automatic axes, you might want to add some minor tick marks in between the major tick marks on your axes. This is covered in section 3.3.2 but, briefly, the 'Hmisc' package (Harrell et al. 2021) contains the minor.tick function which is structured:

```
minor.tick(ny=, nx=, tick.ratio=)
```

where ny and nx specify the number of minor tick marks to place between y-axis and x-axis major tick marks, respectively. tick.ratio is a value that gives the size of the required minor tick marks relative to the major ones. minor.tick is a function that should be run after your graph has been produced; we used it to add minor tick marks to both axes in Figure 7.6b.

7.5.2 Drawing new axes

Sometimes the axes R automatically produces will not do what you want them to, and in such cases, where you want complete control over axis spacing and tick marks, then drawing your own axis/axes is the easiest option. This is simple to do.

First, you need to produce your graph with defined axis limits (using xlim and ylim, as mentioned in section 7.5.1) but with the tick marks and labels for whichever axis/axes you want to redraw suppressed; this is achieved by including xaxt="n" and/or yaxt="n".

Next, you can use the axis function to draw on your own custom axis:

```
axis(side=, at=, labels=, tck=)
```

The first main argument to include in your axis code is which side of the plot it is to be drawn on. As with plot boundaries (discussed in section 7.2.1) and mtext (discussed in section 7.3.5), the possible values for side (1–4) refer to the bottom, left, top, and right sides of the graph, respectively. With at, you then list the values at which you want tick marks positioned along the axis, bearing in mind the minimum and maximum of the axis limits you defined; you can use the function seq to do this concisely if your ticks are to be evenly spaced (see Further Reading). Using labels, you then create a corresponding list, of the same length as at, of the labels you want positioned at each of the tick marks (again, seq could be used to list evenly spaced values). If, for any reason, you wanted some of the tick marks to be left unlabelled, you could fill in their space in the list with " ". You could then also specify the length of the tick marks you want, as a fraction of the plotting region, with `tck`; negative numbers plot outside the graph while positive numbers can be used to plot inside the graph.

Further optional features for custom axes include colour (see section 7.3.1 for details), line type (see section 7.3.4 for details), tick mark orientation (see section 7.5.1 for use of `las`), and the argument `pos`, which is where you can specify the coordinate at which the axis line is to be drawn (which is only really necessary if you do not want your axis positioned at the base or side of the plot).

You might use `axis` first to position any major tick marks and then run a second `axis` command for the same plot side to add minor ticks with different values assigned to the customizable arguments. When preparing minor tick marks, you might be more likely to want to leave them unlabelled, and here the `rep` function can be useful in creating a list of blank `labels` of the required length; we do so for Figure 7.6c and in section 8.5 when preparing logarithmic axes (see also Further Reading).

In Figure 7.6, we have plotted some simple data generated in R on three plots. The first plot (Figure 7.6a) plotted the data using the automatic axes R produced for this data given no axis customization code. The second plot (Figure 7.6b) shows the data plotted against axes that have had their limits modified within the code for the plot itself, and then minor tick marks added

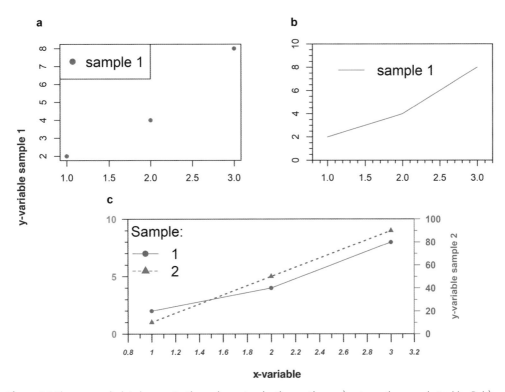

Figure 7.6 Three-panel plot demonstrating axis customization options: a) automatic axes plotted by R, b) modified axes, with set limits and minor tick marks added, c) entirely customized axes, with defined tick mark spacing and sizes, and tick label sizes, font styles, orientations, and colours. The code for the figure (provided in section 7.5.2) also includes many of the graphical parameter and design feature customizations covered previously in this chapter.

using the `minor.tick` function from the 'Hmisc' package (discussed in section 7.5.1). The third plot (Figure 7.6c) shows two samples of data plotted on a graph with three separately added custom axes. The original plot was created with defined axis limits but suppressed tick marks. Customized x- and y-axes were then added, with newly defined tick mark spacing and sizes, and tick label sizes, font styles, orientations, and colours. We provide the code for the entire figure below, in which you should also be able to identify how we:

- Arranged the three plots as a single figure in the plotting region, and plotted two samples together on Figure 7.6c (see section 7.2 for details on modifying plotting parameters).
- Customized various design features, including colours (see section 7.3.1), font (see section 7.3.2), and line types (see section 7.3.4).
- Created shared axis labels and consistently positioned individual panel labels (as discussed in section 7.3.5).
- Added legends which vary in their design (see section 7.4 for legend customization details).

Figure 7.6 code:

```
xvar<-c(1,2,3)
yvarsam1<-c(2,4,8)
yvarsam2<-c(10,50,90)
par(mfrow=c(2,2))
par(mar=c(3.5, 4.1, 3.1, 2.1))
par(oma=c(0,1,0,0))
## Panel a:
plot(xvar,yvarsam1,pch=16,col="red",xlab="",ylab="")
legend("topleft",pch=16,cex=1.5,col="red",legend="sample 1")
mtext("a.", side=3,line=1,adj=-0.1,cex=1.5,font=2)
## Panel b:
plot(xvar,yvarsam1,xlim=c(0.8,3.2),ylim=c(0,10),xaxs="i",
  yaxs="i", pch=16,col="white",xlab="",ylab="")
lines(xvar,yvarsam1,col="red")
legend("topleft",bty="n",inset=0.075,cex=1.5,
  lty=1,col="red",legend="sample 1")
mtext("b.", side=3,line=1,adj=-0.1,cex=1.5,font=2)
library(Hmisc)
minor.tick(ny=4, nx=5, tick.ratio=0.5)
## Panel c:
par(fig=c(0.1,0.9,0,0.5),new=TRUE)
par(oma=c(1,3,0,3))
par(mar=c(4.1, 4.1, 2.1, 2.1))
plot(xvar,yvarsam1,xlim=c(0.8,3.2),ylim=c(0,10),xaxs="i",
  yaxs="i", xaxt="n",yaxt="n",pch=16,col="red",xlab="",ylab="")
lines(xvar,yvarsam1,col="red")
axis(side=1,at=c(seq(0.8,3.2,0.2)),
  labels=c(seq(0.8,3.2,0.2)),tck=-0.07,cex.axis=0.8,font.axis=4,
  col.axis="purple4",las=1)
axis(side=1,at=c(seq(0.9,3.1,0.2)),
  labels=c(rep("",12)),tck=-0.035)
```

```
axis(side=2,at=c(0,5,10),labels=c(0,5,10),tck=-0.05,
  cex.axis=1, font.axis=2,col.axis="forestgreen",las=2)
axis(side=2,at=c(seq(0,10,1)),labels=c(rep("",11)),tck=-0.025)
par(new=TRUE)
plot(xvar,yvarsam2,xlim=c(0.8,3.2),ylim=c(0,100),xaxs="i",
  yaxs="i", xaxt="n",yaxt="n",pch=17,col="royalblue",xlab="",
  ylab="")
lines(xvar,yvarsam2,lty=3,lwd=2,col="royalblue")
axis(side=4,col="royalblue",at=c(seq(0,100,20)),labels=
  c(seq(0,100,20)),tck=-0.05,cex.axis=1,font.axis=2,col.
  axis="royalblue",las=2)
axis(side=4,col="royalblue",at=c(seq(0,100,10)),labels=
  c(rep("",11)),tck=-0.025)
legend("topleft",bty="n",cex=1.5,lty=c(1,2),col=c("red",
  "royalblue"),pch=c(16,17),legend=c("1","2"),title="Sample:")
mtext("y-variable sample 2", side=4,line=3,cex=0.8,
  col="royalblue",font=2,xpd=TRUE)
mtext("c.", side=3,line=1,adj=-0.1,cex=1.5,font=2)
## Add shared axes on blank dummy plot
par(fig=c(0,1,0,1),new=TRUE)
mtext("x-variable", side=1,line=4,cex=2,font=2,xpd=TRUE)
mtext("y-variable sample 1", side=2,line=5,cex=2,font=2,
  xpd=TRUE)
## Return to defaults
par(mfrow=c(1,1))
par(oma=c(0,0,0,0))
par(mar=c(5.1, 4.1, 4.1, 2.1))
```

💡 Key point

Axes are critical to a good graph. Often R's defaults will produce a good set of axes for you. But if you would like to make any change in your axes, a little exploration will show you how to get the look you want.

≋ Chapter Summary

- R is a hugely flexible graphics package, and you can modify practically any aspect of a figure you can imagine.
- By controlling the way your figures are displayed, you can improve their functionality and/or aesthetic appeal.
- Design components of figures, such as colour, font, points, and lines, can be customized to creative informative and visually appealing figures.
- Legends are important features that allow easy interpretation when multiple colours/points/lines are used in a single figure.
- Axes can be entirely controlled in R, to help communicate the data in as clear a way as possible.

Online Resources

The following online resource is available for this chapter at www.oup.com/he/humphreys-obp1e:

- R script for chapter 7

Further Reading

- The blog 'Mastering R Plot – Part 3: Outer Margins' is the third in a series on customizing your plots, and explains how to alter the space around your charts to give more flexibility when adding legends and axis labels:

  https://www.r-bloggers.com/2016/03/mastering-r-plot-part-3-outer-margins/

- 'Combining Plots' shows you how to create multi-panel figures, with a number of plots in one overall graph:

  https://www.statmethods.net/advgraphs/layout.html

- 'R Function of the Day: layout, par(mfrow)' gives the code to produce and customize a multi-panel figure:

  http://rfunction.com/archives/1538

- 'Arranging Plots with par(mfrow) and layout()', a section from *YaRrr. The Pirate's Guide to R*, looks at different ways to produce complex multi-plot layouts:

  https://bookdown.org/ndphillips/YaRrr/arranging-plots-with-parmfrow-and-layout.html

- 'Tips for Laying Out Plots in R' gives useful further details on creating complex layouts in R:

  https://www.staff.ncl.ac.uk/stephen.juggins/data/Iowa2007/PlottingTips.pdf

- 'Graphical Parameters of R {graphics} Package' is a comprehensive document of examples of use for each parameter of the **par** function in base R:

  http://rstudio-pubs-static.s3.amazonaws.com/315576_85cccd774c29428ba46969316cbc76c0.html

- 'Tired of Using Helvetica in Your R Graphics? Here's How to Use the Fonts You Like' explains how to import non-standard fonts, particularly for use in PDF or postscript files:

  http://zevross.com/blog/2014/07/30/tired-of-using-helvetica-in-your-r-graphics-heres-how-to-use-the-fonts-you-like-2/

- 'Changing the Font of R Base Graphic Plots' gives a way to change the font when working in Mac OS:

  http://rcrastinate.blogspot.com/2015/08/changing-font-of-r-base-graphic-plots.html

- 'Cookbook for R: Fonts' offers another way to change the font in your charts, using **geom_text** or **annotate** in **ggplot2**:

  http://www.cookbook-r.com/Graphs/Fonts/

- 'points: Add Points to a Plot' explains how to use the **points** function:

  https://www.rdocumentation.org/packages/graphics/versions/3.6.2/topics/points

- 'Quick Guide to Line Types (lty) in R' shows you how to create your own line type, line end styles, and line join styles:
  https://www.benjaminbell.co.uk/2018/02/quick-guide-to-line-types-lty-in-r.html
- 'Line Plots—R Base Graphs' takes you through the code to create line plots, including multiple lines:
  http://www.sthda.com/english/wiki/line-plots-r-base-graphs
- 'Paste Function in R':
  https://www.datasciencemadesimple.com/paste-function-in-r/
- 'Round Function in R':
  https://www.datasciencemadesimple.com/round-function-in-r/
- 'Add Custom Tick Mark Labels to a Plot in R Software':
  http://www.sthda.com/english/wiki/add-custom-tick-mark-labels-to-a-plot-in-r-software
- 'Axes and Text':
  https://www.statmethods.net/advgraphs/axes.html
- 'R seq Function':
  http://www.endmemo.com/r/seq.php
- 'R rep Function':
  http://www.endmemo.com/r/rep.php

8 CUSTOMIZING EVERYTHING USING R: MORE SPECIALIST

Learning objectives

By the end of this chapter you should be able to:

- Produce figures with inset plots.
- Add frames, images, background images, arrows, symbols, and shapes to figures.
- Produce figures with logarithmic axes.
- Annotate images and geographic maps within the plotting area of R.
- Appreciate the flexibility R offers in terms of customization.

8.1 Introduction

This is the second of two chapters in this book dedicated to customization of graphs. This is in large part due to the wealth of customization R offers when graphing, and the fact that most features are adaptable to any of the main graph types covered in this book. The earlier customization chapter covered essential, day-to-day pieces of code that should all be helpful to you repeatedly, but here we turn to more specialist aspects of customization in R. While this chapter still essentially provides walk-throughs of R code using specific data examples, its main purpose is to provide generic and editable code for a range of plotting and design features. We encourage you simply to dip into the section that details a feature of interest to you, read through (and perhaps run through the accompanying script) to understand the required code in the context we provide, and then adapt the code to your needs when applying it to your own graph.

The chapter is shaped around two main examples. The first of these we previously presented as a bar chart with an inset earlier in the book, but here we explain how it was produced. We then use this figure as a base from which to explain the methods of framing plots, adding images, and adding background images to figures in R. The second example is fictitious exponential data that we generate in the accompanying script, with which we first demonstrate how to set up and customize logarithmic axes. The resulting scatterplot is then used as a base to which we add arrows, symbols, and shapes using

R customization code. In the final section 8.7, we also briefly cover how to annotate images or geographic maps in the R plotting space using tricks covered earlier in the chapter.

 Key point

Unless your graph type is very specialist, there will be a way to produce it, and produce it exactly the way you want, in R.

8.2   Inset a plot within a figure

In sections 3.7 and 6.6 we discussed the use of inset plots as an effective way to start y-axes at non-zero values in a figure while reducing the risk of this misleading readers. Here, we will walk through the code used to produce SA 3.2 Figure A, which presents both a 'zoomed-in' view and 'big picture' inset of biennial Canadian per capita greenhouse gas (GHG) emissions from 2005 to 2017 (Environment and Climate Change Canada 2020); the data is available to download as the file 'canada_emissions.csv' in the online resources.

Step 1: First, load in the data set and have a look at the sort of data we have:

```
ghg <- read.table(file.choose(), header = T, sep = ",")
View(ghg)
```

The first column lists years ('year') and the second column lists the metric tonnes of CO_2 equivalent per capita ('emissions'). The values for the emissions we'll be plotting up the y-axis are continuous, quantitative data (see section 1.3 and the online resource 'Data types' from chapter 1 for details) and, as we mention in chapters 3 and 5 (see also the Further Reading in chapter 5), we would not typically recommend the use of bar charts for this sort of data. However, we will demonstrate here how to produce a bar chart with an inset to match SA 3.2 Figure A, as the process is the same if you wanted to add inset plots to bar charts for qualitative data.

Looking at the data, we can see that the change in values of GHG emissions only spans around 3 metric tonnes (from 19.5 to 22.6). However, depending on the purpose of our figure, it might be worth highlighting any decline in emissions across the time frame, given that this could herald important environmental benefits. For this reason, we will present a zoomed-in version of the data as our main figure, with a non-zero y-axis starting at 19 metric tonnes, and include an inset of the emissions in context (that is, with a y-axis starting at zero).

Step 2: Before we plot our main figure, we need to use the **par(fig)** graphic parameter to set out limits for our main figure. We touched on this code in section 7.2.3 but, to briefly recap, **fig** lets us precisely control the location of a figure within the plotting area—we just need to provide coordinates of plot corners in the form (x1, x2, y1, y2). Using **fig** in this context allows us to create coordinates from which we can base the placement of the inset plot later. To tell R that we want the first figure we plot (which will be our zoomed-in bar chart) to fill the entire device space we run the code:

```
par(fig = c(0,1,0,1))
```

This is actually what R would normally assume for plots, i.e. that the figure should fill the whole plot area (with both x and y running from 0 to 1). However, it is useful to include **fig** ahead of our main plot to give context to the coordinates we will specify when placing the inset.

Step 3: We next plot our zoomed-in bar chart, with a non-zero y-axis that runs from 19 to 23 metric tonnes, which will fill the plotting area. We assign the name '**canadabar**' to this version of the plotted data, to differentiate the versions and to make it easier to add on extra features. As this is a bar chart we are giving a non-zero y-axis to, we have to include the argument **xpd=FALSE** to prevent the bars extending outside (here, below) the plotting area—have a go at running the command without the **xpd=FALSE** argument to see its effect for yourself. We also give our main bar chart a clear y-axis baseline at 19 metric tonnes (using **abline**; see section 7.3.4 for more on customizing lines) and plot our own tidy x-axis with tick marks leading from the middle of each bar (see section 7.5.2 for a recap of how to draw new axes):

```
canadabar<-barplot(ghg$emissions~ghg$year, cex.names=0.8,
  ylim=c(19,23),yaxs = "i",xaxt="n",
  ylab=expression(GHG~Capita~(t~CO[2]~eq/capita)),xlab="",
  col="seagreen1",xpd=FALSE)
abline(h=19)
axis(side=1, at = canadabar, labels=ghg$year,las=1)
```

For our x-axis placement remember that, relative to the plot, **side** places an axis: 1=below, 2=left, 3=above, or 4=right. For the **at** argument, we can just use our name '**canadabar**', and R will know to position x-axis major tick marks centrally below each bar of this chart. You might also notice that the y-axis label argument (**ylab**) looks pretty busy. This is because rather than just print the text 'CO2' we have decided to present the formula for carbon dioxide with the '2' as a subscript, to get 'CO_2' in our y-axis label. We do this using the function **expression**, then listing our label text with the ~ symbol as a space character, and isolating the '2' to be subscripted in the middle by enclosing it in square brackets [ ]. **expression** tells R to print anything that is contained in square brackets as subscript.

You may recall our use of **expression** in section 6.4.1 to include the degree symbol for temperature, but in that case we used **paste** instead of punctuation to separate out components of code. You are likely to come across both approaches elsewhere, but the one you adopt should simply be the one that feels best to you (see Scientific Approach 8.1 for more details on these two approaches to constructing **expression** commands). We will encounter more advice and complex examples of the use of **expression** and formulae (and **paste**) in section 8.6.

Step 4: Now to position our inset plot. The top-right of the main bar chart looks to have the most space available for an inset, as the bars there don't reach as far up the y-axis. Bearing in mind that our main plot fills the plot space with the x-coordinates running 0–1 and the y-coordinates running 0–1, we need to think about which coordinates within these ranges would be good for the placement of our inset. In choosing these coordinates it can sometimes take a bit of trial and error to get the proportions of your insets right, and the associated text legible, but the **fig** code below includes some values that we think work well for this example:

```
par(fig = c(0.4,1, 0.35, 1), new = TRUE)
```

Scientific Approach 8.1
Constructing multicomponent expression commands in R

The **expression** function in R allows you to incorporate special characters, symbols, or formatting in your text, plot labels, or other graphical features. For example, subscript can be displayed using square brackets [ ], superscript can be displayed using the caret symbol ^, and mathematical symbols not included on regular keyboards can be printed by calling their specific names. However, in cases where such specialist use is required as part of a larger, multicomponent text or symbol addition, the code used within **expression** needs to be built up in sections. There are two main alternative approaches with which you can construct expressions in R.

The first involves the use of the **paste** function. The **paste** function should be positioned immediately after the **expression** function, text should be contained within quotation marks " ", commas should be used to separate components of the code, and any spaces to occur within sections of text or between components should be contained within quotation marks " ". As an example, take the following line of code:

```
expression(paste("Subscript"[subscript],
" ","Superscript"^superscript,"        ",
"plain text"," ",gamma))
```

Using **paste**, this code would print (in order) the word 'Subscript' with a subscript of its own that reads 'subscript', a single space, the word 'Superscript' with a superscript of its own that reads 'superscript', five spaces, the words 'plain text' with no special formatting applied, another single space, and then the Greek letter gamma. See section 8.6 for further examples where we use this approach.

The main alternative to using **paste** is to use the function **expression** on its own and use additional punctuation. In this case, the ~ symbol acts as a space character and the asterisk * acts as a connector that allows you to join together code components (e.g. if presenting text with normal formatting alongside text or symbols with special formatting like italics) and/or can act as a 'stop' command for any special formatting that was previously applied in the command, such as superscript. Quotation marks " " are then only used to enclose items that would otherwise be treated as a special or instructional character (such as ~ or *). The following line of code would produce exactly the same output as the line using **paste** above:

```
expression(Subscript[subscript]~Superscript^
superscript~~~~~plain~text~gamma)
```

Note the lack of quotation marks and extensive use of the space character (especially for the five-space gap ahead of 'plain text'). See section 8.7 for an example of asterisks being used to connect specially formatted text or characters (e.g. italics) with normal text or components.

Have a go at running the code provided for this Scientific Approach in the online resource 'R script for chapter 8' to see how both approaches give identical outputs in this example. Often, the '**expression** alone' approach can be the briefer way to construct code, but you may find it more helpful to think of code as being structured in chunks contained by quotation marks (as with **paste**). Occasionally, there may be something particular you want to achieve for which there is detailed advice for one approach but not the other available online—one way or another, with a bit of perseverance you will always find a solution that works. See the Further Reading for a few additional pointers regarding **expression**.

Notice that we also include the **new=TRUE** argument within our overall **par** code—this allows us to plot our inset on top of the existing figure, in the same way that it has allowed us to plot multiple samples of data onto the same plot previously (see chapters 5 and 6).

Step 5: Finally we plot our inset bar chart (with y-axis running from 0) into the top-right of the existing plot at the coordinates defined above. We name this version of the plotted data '**fullbar**'. Again, we add a horizontal y-axis baseline (though here at 0) and a tidy, customized x-axis (which uses the centre of the bars in 'fullbar' to position tick marks):

```
fullbar<-barplot(ghg$emissions~ghg$year, cex.axis=0.7,
  ylim=c(0,25),yaxs = "i",
  ylab="", xlab="", xaxt="n",
  col="seagreen1")
abline(h=0)
axis(side=1, at = fullbar, labels=ghg$year,las=2,
  cex.axis=0.7,tck=-0.05,hadj=0.6)
```

In order to fit the x-axis tick marks and text neatly in our small inset, you will notice that we have played around with the axis text orientation (**las**), font size (**cex.axis**), tick mark length (**tck**), and positioning of the tick labels (**hadj**, or 'horizontal adjustment'). Again, settling on these values can take a bit of trial and error (and see section 7.5.2 for more on axis customization).

In Figure 8.1, we have produced an exact replica of SA 3.2 Figure A from chapter 3. However, the use of insets is not restricted to bar charts, as with other figure types you might also initially face a choice between misrepresenting data (by changing the y-axis) or miss important patterns (by forcing the y-axis to start at 0). For example, in section 6.6 we touched on the use of non-zero y-axes in scatterplots—another case where insets can be used to give context while important patterns in your data are highlighted. In fact, as the Canadian emissions data shows change in a continuous variable over time (another continuous variable), a scatterplot—specifically a time series—is the figure type we would recommend for this sort of data. So, we will now present the same data as a time series with an inset. Try running the code:

```
par(fig = c(0,1,0,1))
plot(ghg$emissions~ghg$year, xlim=c(2005,2017),
  xaxs = "i",ylim=c(19,23), yaxs = "i", xlab="Year",
```

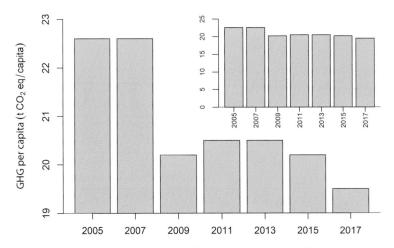

Figure 8.1 A replica of SA 3.2 Figure A, chapter 3. Biennial Canadian per capita greenhouse gas (GHG) emissions from 2005 to 2017 (Environment and Climate Change Canada 2020).

```
  ylab=expression(GHG~Capita~(t~CO[2]~eq/capita)),
  pch=16, col="seagreen1",xaxt="n")
axis(side=1, at = seq(2005,2017,2), labels=ghg$year,las=1)
sortedvals <- ghg[order(ghg$year),]
lines(sortedvals$year,sortedvals$emissions,lty=1,lwd=2,
  col="forestgreen")
par(fig = c(0.4,0.95, 0.35, 0.95), new = TRUE)
plot(ghg$emissions~ghg$year, cex.axis=0.7,
  xlim=c(2005,2017), xaxs = "i",ylim=c(0,25),
  yaxs = "i", xlab="", ylab="", pch=16, col="seagreen1",
  xaxt="n")
axis(side=1, at = seq(2005,2017,2), labels=ghg$year,las=2,
  cex.axis=0.7,tck=-0.04,hadj=0.6)
lines(sortedvals$year,sortedvals$emissions,lty=1,lwd=2,
  col="forestgreen")
```

The time series-specific code used above can be found in section 6.4, but you will recognize all of the code required to add an inset plot as being exactly the same as in our bar chart example earlier, with the exception of a few slight tweaks of values. The reason for the slightly different numbers in our **par(fig)** code for the inset plot becomes clearer when we look at the resulting Figure 8.2.

Figure 8.2 shows the same data as Figure 8.1 but in what many would consider a more appropriate format for continuous, quantitative data; just as in Figure 8.1, the main plot shows the significant decline in emissions over time, while the inset presents that decline in a broader context. As noted above, when placing the inset plot, the coordinates used did not reach to 1 on either the x- or y-axes as they did for Figure 8.1. This is because the **plot** function in R automatically adds a frame around plots (that is the rectangular box enclosing each plot), whereas the **barplot** function does not—the coordinates needed to be

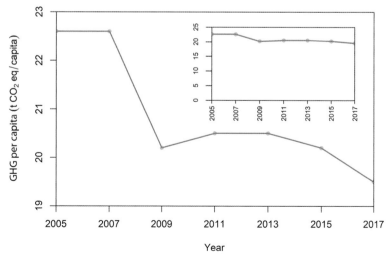

Figure 8.2 A time series showing biennial Canadian per capita greenhouse gas (GHG) emissions from 2005 to 2017 (Environment and Climate Change Canada 2020).

edited to prevent the axes of the inset running into the frame of the larger plot. If you like the look of the frames around the plots in Figure 8.2, we will next show you in section 8.3 how you could add a frame around bar charts and other figures where frames are not the default in R.

 Key point

Insets are handy in lots of ways, and you will often see them as a good way to include a useful but simple image that doesn't really require a figure or even a figure panel all to itself.

8.3   Adding and removing frames from plots

There are two useful positions we can add frames (or boxes) around charts in R: i) around a plot (just encasing the plotting area itself), or ii) around the overall figure (including the axis labels and title, if applicable). In R, the functions for some figure types produce a default frame around the plotting area (e.g. box-plots and scatterplots), but other plots (e.g. bar charts and histograms) do not. None of the figure types are produced with a frame around the overall figure as default. The code to add a frame to either location is really simple, and here we'll demonstrate how to do both on the bar chart from section 8.2.

Step 1: First, run the code used to create Figure 8.1, the bar chart with an inset (easily done using the online resource 'R script for chapter 8'), only this time when placing the inset plot we will specify:

```
par(fig = c(0.4,0.95, 0.35, 0.95), new = TRUE)
```

These are the coordinates used for the time series inset in Figure 8.2, and they will position the inset a little way in from the very edges of the main plot (allowing space for us to add a distinct frame around the inset).

Step 2: Now, again using **par(fig)** to remind R of the plotted position of our inset, we can encase the subplot in a frame using **box**:

```
par(fig = c(0.4,0.95, 0.35, 0.95))
box(which = "plot", lty = 1)
```

box is the function that draws the frame, **which** specifies which of the positions we want the box around (here **"plot"**, just the plotting area), and—as ever—line type **lty** (and/or line width **lwd** or colour **col**) could be edited. You should see that there is now a box around the inset bar chart connecting the x- and y-axes, just as happens to scatterplots by default in R.

Step 3: We can then do exactly the same for our main plot, by specifying its coordinates using **par(fig)** first:

```
par(fig = c(0,1, 0, 1))
box(which = "plot", lty = 1)
```

Step 4: And now, to show the other main position where frames can be added using **box**, we will draw a slightly thicker, dashed blue line around the entire figure (encompassing the axis labels). Again, we need to specify the **par(fig)** to tell R that we want the whole of the main plot rather than just the inset plot framed:

```
par(fig = c(0,1,0,1))
box(which = "figure", lty = 2, lwd=3,col="blue")
```

In Figure 8.3, you can see the thin black frames around each plotted bar chart and the thicker, dashed blue frame around the entire figure. We have seen here how the **box** function relies on coordinates given by **par(fig)** to position frames around the very margins of plots. However, the coordinates that span the values within those plot margins are also very useful when customizing figures in R, and are used when positioning images (discussed next in section 8.4) and adding features such as symbols and shapes (as discussed later in section 8.6).

Before we move on to adding images, we will briefly explain how removing default frames from certain figure types is even easier than adding frames. To remove the frame R automatically surrounds the plotting area of scatterplots and boxplots with, you simply need to include the argument **frame=FALSE** in the **plot** or **boxplot** function and run the full commands for the figure. Removing the frame from scatterplots leaves just the x- and y-axes on the bottom and left sides of the plot. To demonstrate, in the online resource 'R script for chapter 8' we provide code which produces a version of the time series in Figure 8.2 but with the **frame=FALSE** argument included in both the **plot** function for the main plot and the inset plot. Removing the frame from a boxplot leaves you with just a y-axis. The 'R script for chapter 8' also includes code for a basic boxplot showing descriptive statistics for just the GHG emissions, for which we use **frame=FALSE** to remove three of the default frame sides around the plotting area. Most of the time, you will not need to remove the default frames produced with certain figure types—they can help with readers' spatial awareness and make interpretation easier. However, for some occasions (for example, presentations) you might decide that removing the top and right sides of the frame around scatterplots, or the top, right, and bottom sides of the frame around boxplots, will look more minimalist and visually appealing.

8.4   Adding images and background images

On most occasions where you are considering adding images to a figure, the image itself will not add to a viewer's understanding of the data—rather, it is there as an appealing embellishment. For this reason, figures published in academic journals rarely contain images, but photo or cartoon visuals are often found in figures designed for presentations or literature targeted at a less specialist audience. To include images in a figure that you are creating in R, you will first need to have found the images you want to include and saved them to your computer. Most search engines will have a setting, sometimes under 'Advanced Settings', that will allow you to narrow down image search results by file type—here we will show you how to include both PNG and JPEG image files in your figures.

Imagine that, for the data set on Canadian GHG emissions, we also have data on per capita meat consumption in Canada and that the decline in emissions occurred concurrently with changes in meat preferences. In our imaginary data, beef was the most popular meat in years 2005 and 2007, pork was the most popular meat in years 2011 and 2013, and chicken was the most popular meat in the years 2009, 2015, and 2017. We know that it's pretty silly to imagine that preferences amongst meats would be linked so tightly to overall GHG emissions. Further, in the real world, Canadians have actually consumed more chicken than beef or pork every year between 1998 and 2019 (Statista 2020). However, buy into our imaginary scenario for a bit and, for the purposes of illustration, we'll add images associated with our imaginary data.

Step 1: The images we provide in the online resources for this example are PNG files, so we first have to install and activate the package '**png**' (Urbanek 2013):

```
install.packages("png")
library(png)
```

You may also commonly have images saved as JPEG files, but the code to add JPEGs to plots is very similar. We will add a JPEG image as a background to our plot later in this section to demonstrate.

Step 2: Next, the images we want to include each need to be read in and given a name (though, first remember to save them to your device). The PNG images we provide as online resources are called 'cow', 'pig' and 'chicken', but running the code below will associate them instead with the labels '**beef**', '**pork**' and '**chicken**' respectively. For each line of code, you will need to select the appropriate PNG file from your device's folders:

```
beef<-readPNG(file.choose())
pork<-readPNG(file.choose())
chicken<-readPNG(file.choose())
```

Step 3: Now we have two options. If you have kept your R session running and most recently ran the commands to encase the plotting area and the whole figure with boxes, as covered in the steps of section 8.3, you will already have a bar chart with inset and frames in your plot window.

We could technically just add our images onto the bars that we drew before. But, because we used **par(fig)** to create new plotting regions for the inset and for the addition of frames, we have overwritten R's memory of the main plot's axis limits. This means that we would not have any axes to help us figure out what coordinates are necessary to place images effectively. It is possible to place images despite this, through trial and error, and you can see our attempts at this in the online resource 'R script for chapter 8'. However, the beauty of producing figures with R is that the code and figures we create are entirely reproducible. Therefore, what we will do instead, to make our lives a little easier, is to run each of the two plots in turn and add images to each separately before drawing our frames back on. So we now first replot the main bar chart:

```
par(fig = c(0,1,0,1))
canadabar<-barplot(ghg$emissions~ghg$year, cex.names=0.8,
  ylim=c(19,23),yaxs = "i",xaxt="n",
  ylab=expression(GHG~Capita~(t~CO[2]~eq/capita)),
  xlab="", col="seagreen1",xpd=FALSE)
abline(h=19)
axis(side=1, at = canadabar, labels=ghg$year,las=1)
```

Step 4: Before we add our inset, then, we will add our meat consumption images to the bases of the bars of our main chart using **rasterImage**. This time we provide the coordinates in the form of (x1, y1, x2, y2). Because we are adding images immediately after plotting our figure, we can base our y-coordinates on the y-axis values. If we were adding images to a figure that also had an intrinsic numerical metric for its x-axis (such as a scatterplot) we would also simply base our x-coordinates on the x-axis values. However, figuring out the x-axis coordinates can be a little tricky where the data on the x-axis doesn't have a clear, continuous scale (such as when adding images to a bar chart). Where quantitative axes aren't present, `rasterImage` requires vectors (or scalars) of x- and/or y-positions—you could have a go at playing with values until you figured out some that worked. Instead, though, we will use a clever shortcut to get R to figure out some of the x-axis placement for us. Because we assigned a name to this bar chart, '`canadabar`', for each image we want to place in line with a bar, we can refer to the central point of that bar by specifying the position of that bar in square brackets after the overall chart name. For example, `canadabar[1]` would be the central x-axis coordinate for the first bar in the bar chart, and `canadabar[5]` would be the central x-axis coordinate for the fifth bar. To set coordinates x1 and x2 (the left and right end points) for our figure placements, then, we need to—respectively—subtract and add on small values to these central coordinates (try a little trial and error to suss out how small). In the code below, we give values that allowed for fairly neat positioning:

```
rasterImage(beef, canadabar[1]-0.48, 19.05, canadabar[1]+0.42,
   19.7)
rasterImage(beef, canadabar[2]-0.48, 19.05, canadabar[2]+0.42,
   19.7)
rasterImage(chicken, canadabar[3]-0.3, 19.05, canadabar[3]
   +0.3, 19.4)
rasterImage(pork, canadabar[4]-0.4, 19.05, canadabar[4]+0.4,
   19.5)
rasterImage(pork, canadabar[5]-0.4, 19.05, canadabar[5]+0.4,
   19.5)
rasterImage(chicken, canadabar[6]-0.3, 19.05, canadabar[6]
   +0.3, 19.4)
rasterImage(chicken, canadabar[7]-0.3, 19.05, canadabar[7]
   +0.3, 19.4)
```

When selecting coordinates for image placement it is important to play about with the numbers until the proportions look right; this is trickier when using axes with no obvious intrinsic scale, but doable with a little trial and error to help you get your bearings. Once you have found values for one image that look right, these can often be used as a guide for the placement and sizing of further images. For example, in our code above we kept the x-axis ranges (those values subtracted and added on to the central x-axis coordinate of each bar) of each image type consistent ('`beef`' = 0.9, '`pork`' = 0.8, '`chicken`' = 0.6) to ensure that images of the same animals were all the same width. Note that while most of the time we would subtract and add on equal values either side of a bar's central coordinate to position images centrally, for the '`beef`' images these values are a little unbalanced—this is simply because we thought the image of the cow fitted the space within bars better with a little more space to the right. For another example of using the bars of a 'named' bar chart to add images using `rasterImage`, check out the online resource 'R script for chapter 1', where we show you how we added evenly sized images of flags to bar chart A in Figure 1.3.

Step 5: Once happy with the placement of images on the main plot, we can then run the code to produce our inset plot '`fullbar`' again:

```
par(fig = c(0.4,0.95, 0.35, 0.95), new = TRUE)
fullbar<-barplot(ghg$emissions, cex.axis=0.7,
```

```
  ylim=c(0,25),yaxs = "i",
  ylab="", xlab="", xaxt="n",
  col="seagreen1")
abline(h=0)
axis(side=1, at = fullbar, labels=ghg$year,las=2,
  cex.axis=0.7,tck=-0.05,hadj=0.6)
```

Step 6: And, just as before, we can add our meat consumption images to the corresponding bars of 'fullbar':

```
rasterImage(beef, fullbar[1]-0.48, 0.05, fullbar[1]+0.42, 4.7)
rasterImage(beef, fullbar[2]-0.48, 0.05, fullbar[2]+0.42, 4.7)
rasterImage(chicken, fullbar[3]-0.3, 0.05, fullbar[3]+0.3,
  4.4)
rasterImage(pork, fullbar[4]-0.4, 0.05, fullbar[4]+0.4, 4.6)
rasterImage(pork, fullbar[5]-0.4, 0.05, fullbar[5]+0.4, 4.6)
rasterImage(chicken, fullbar[6]-0.3, 0.05, fullbar[6]+0.3,
  4.4)
rasterImage(chicken, fullbar[7]-0.3, 0.05, fullbar[7]+0.3,
  4.4)
```

You might have spotted that the values given either side of the central x-coordinates for each bar are identical to those used for the images in the main plot; this is because the relative spacing of the bars is the same. Consequently, the x-axis ranges of each image type remain consistent. The y-axis coordinate values, though, have changed drastically as our inset has a y-axis running from 0–25 GHG/capita instead of the 19–23 GHG/capita y-axis of the main plot; remember from above that coordinates are given in the order (x1, y1, x2, y2). Because the images of the inset are much smaller, we have also reduced the difference in image height (that is the y-axis ranges stated as coordinates) between the meat types, so that the 'chicken' images are still identifiable. Again, we draw your attention to the online resource 'R script for chapter 1', where we provide another example of using the bars of a 'named' bar chart to add images when producing bar chart A in Figure 1.3.

Note: You can also add images to figures outside of the plotting area (that is, outside the limits of the x- and y-axes) by preceding the **rasterImage** function with the code **par(xpd=TRUE)**.

If desired, you could now add the frames back around the inset, the main plot, and the overall figure using the same code as covered in section 8.3:

```
par(fig = c(0.4,0.95, 0.35, 0.95))
box(which = "plot", lty = 1)
par(fig = c(0,1, 0, 1))
box(which = "plot", lty = 1)
par(fig = c(0,1,0,1))
box(which = "figure", lty = 2, lwd=3,col="blue")
```

In Figure 8.3, you can see where the images added to both the main plot and inset should be plotted when the script for this chapter is run. If these images of animals did actually correspond with accurate and useful data (for example, if those meat preferences were real and found to be related to the decreasing GHG emissions), images might be a useful visual aid to include in a figure for a presentation or information leaflet. However, for an academic piece of work, we would advise you to consider instead using different colours for bars representing years where different meats were the most popular (and including a legend and/or a detailed figure caption to explain how the different colours correspond to differences of meaning).

As mentioned earlier, as well as the PNG files we have just added to our bar chart, JPEG image files can also be easily added to plots (though note that PNG files usually retain image quality better than JPEGs when resized/compressed). We will now show you how to do this in the slightly different context of adding a background image to the entire plot; though JPEGs could be added to plots in the same way we have just used images and, conversely, PNG files could be used as background images. To add an image as a background, we first need to create a blank plotting area. The background image will then be added and we will overlay it with our existing customized bar chart code.

Step 1: Just as with PNG files, we first need to install and activate a package specifically designed to load in JPEG files, '**jpeg**' (Urbanek 2019):

```
install.packages("jpeg")
library(jpeg)
```

Step 2: The image we provide in the online resources to include as a background here is simply of the Canadian flag. So we next need to read in the image file 'Flag-Canada' and give it a name within R. As before, you will first need to save the image to your device and then the code below will bring up your device's folders for you to select it from:

```
canadaflag <- readJPEG(file.choose())
```

Step 3: Similarly to when we set up blank plots ahead of adding grid lines in chapter 6, here we need to set up a blank plot area with no plot and no axes for the image to sit within. Because we have previously been looking at two different plotting areas (the main plot and the inset) we will first specify that we want the plot to fill the available plotting area using **par(fig)**, and then run the code for a blank plot:

```
par(fig = c(0,1, 0, 1))
plot(1:2, type="n", xaxt="n", yaxt="n",xlab="", ylab="")
```

Here our values of '1:2' are pretty arbitrary, as we are hiding the axes anyway, and using the **type="n"** gets us a blank canvas (in chapter 6 we produced blank plots via a slightly different method, but there are often multiple ways the same outcome can be achieved using R).

Note that we need to specify blank x- and y-labels so that R does not plot some unhelpful axis names that will mess up the names we want when we lay our actual labels on top.

Step 4: We now extract the plot information, naming it '**lim**' as it gives us the limits of the plotting area:

```
lim <- par()
```

Step 5: Using **lim**, our plot limits, we can now fill the plot with our background image, using **rasterImage** again:

```
rasterImage(canadaflag, lim$usr[1], lim$usr[3], lim$usr[2],
lim$usr[4])
```

You don't need to get too hung up on the details of how this bit of code works. `usr` is related to `par` (the multifaceted go-to whenever changing the formatting/layout of plots and their margins, covered more in section 7.2) and concerns the upper and lower x and y limits for the plot region. Basically, this code fits our image to the upper and lower limits of each 'side' of our plot.

Step 6: Now, we can simply tell R that we want to add our customized bar chart (with inset, images, and frames) on top of this background, by including `new=TRUE` in the `par` command ahead of plotting the main bar chart '`canadabar`' (where we specified fig as `c(0,1,0,1)`) and running the remaining code (for both '`canadabar`' and the inset chart '`fullbar`').

Altogether, this gives us the resulting Figure 8.3. To our basic code for the bar chart with non-zero y-axis we have added: an inset plot with a y-axis running from zero, frames around both plots and the overall figure, PNG images to both the main and inset plots, and a JPEG image as a background to the main plot. To be clear, we certainly do not recommend that you use all these tricks on any single figure. We have done so here to provide an example of the customization available to you with R, but to our minds Figure 8.3 is too cluttered and the inclusion of too many embellishments (that do not add to the information presented in a meaningful way) obscures the message of the data.

We recommend you follow a 'simple but not too simple' rule of thumb when designing figures—see section 1.7.1 for further discussion of this guideline and the issue of 'chart junk'. However, if you think that adding images to a particular figure will help to communicate its message to its intended audience, see the online resource 'R script for chapter 1' and downloadable images from chapter 1 for a further example (Figure 1.3A).

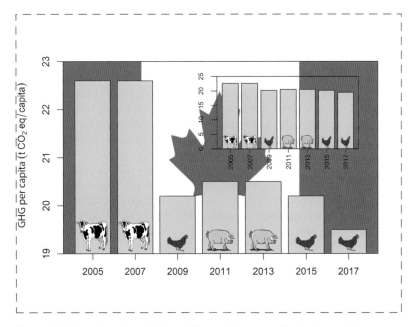

Figure 8.3 A busy bar chart with inset, frames, images, and a background image. The figure shows biennial Canadian per capita greenhouse gas (GHG) emissions (Environment and Climate Change Canada 2020) and the hypothetical most popular meat consumed from 2005 to 2017.

Key point

The ability to add images can be transformative for your figure, but make sure they are actually aiding the reader and not just adding clutter.

8.5   Logarithmic axes

In Bigger Picture 4.1 and section 6.5 we mentioned the use of logarithmic axes as an option for when you need to present quantitative data that spans a very wide range of values in a single figure. Examples of this include exponential data, where the largest measured values in a data set can be hundreds or thousands of times greater than the smallest values. Scatterplots are probably the most common figure type where you will encounter logarithmic axes, so we will here generate a set of fictitious continuous, quantitative data to be plotted as a scatterplot with logarithmic axes.

Step 1: To generate our fictitious data, the code below will create two lists: one of x-values and the other of corresponding y-values:

```
x <- seq(0, 100, 1)
y <- expm1(x)
```

seq tells R we want x to be a series of numbers (in sequence) that runs between 0 and 100 with an interval of 1. **expm1** computes the exponential function of the following number minus 1.

Step 2: We can see that the **expm1** function has successfully generated exponential data if we quickly visualize the resulting values in a simple scatterplot:

```
plot(x, y)
```

Looking at Figure 8.4a, our pattern of exponential growth is evident: the y-values are increasing slowly with x at first and then very rapidly. However, the pattern in the data is not very clear because it's difficult to distinguish the data points and a lot of our data points are squeezed up at the right-hand side of the plot. Using logged x- and y-axes will help us visualize the data better.

Step 3: Let us try plotting instead:

```
plot(x, y, log="xy")
```

Note: Don't worry about the warning messages '1 x value <= 0 omitted from logarithmic plot' and '1 y value <= 0 omitted from logarithmic plot'. These just occur because on a logarithmic scale 0 is minus infinity so R ignores the 0 value when plotting the data.

In this command, the **log** argument tells R whether we want the x- or y-axis or both to be plotted on a logarithmic rather than a linear scale; here we put **"xy"** to make both axes logarithmic. The result of this is clear in Figure 8.4b, where the scales of the axes no longer have even interval values between the major tick marks. This spreads out and makes clearer the most interesting pattern in our data (the rapid increase). Importantly, we are here presenting untransformed data on logarithmic axes—i.e. we did not log the data, we just presented the existing list of values against logarithmic scales.

The other alternative that we have used in Bigger Picture 4.1 in chapter 4 is to present logged data on linearly scaled axes. See also the online resource 'Authors' attempts for chapter 6' and associated 'R script for Authors' attempts chapter 6' for an example of using log-scaled axes with untransformed data and an example of logging the data itself. Our advice, when you feel that your data would benefit from being presented logarithmically, is to choose the option that you and your prospective readers will find easier. Either way, the figure will be more challenging to interpret and you will need to talk readers through the figure more carefully. If you are interested in logging data (perhaps if it spans a broad spread of values and/or is highly skewed) see Bigger Picture 4.1 in chapter 4 for an explanation of the **log** and **log10** functions in R.

While the data in Figure 8.4b looks clearer than it did in Figure 8.4a, our y-axis ticks are still labelled with R's default format for extremely large values (rather than the more familiar '10 to the power' mathematical format), and we might want tick marks labelled on both axes.

Next we will customize our logarithmic axes (see section 7.5 for a reminder of the functions and arguments used to customize axes).

Step 4: Online you might find some complicated functions people have written to edit logarithmic axes, but we find the easiest method is to edit axes in exactly the same way as we have done earlier in this chapter and in previous chapters. That is, we will first plot our figure without the axis, or axes, of interest and then add it, or them, separately. So, here we'll plot our data on x and y log scales, but this time we'll keep details of the axes blank; and we will specify a point style and colour too:

```
plot(x, y, log="xy", xaxt="n",yaxt="n",pch=20,
col="darkmagenta")
```

Step 5: Next, we'll add on the x-axis (using **axis** and specifying **side** 1) with major tick marks specified **at** 1, 10, 50, and 100:

```
axis(side=1, at=c(1,10,50,100),tck=-0.04)
```

Step 6: Just as a reminder of how customizable axes are, we will also add minor tick marks at 5, 25, and 75 manually, by first listing what we want them labelled; we'll leave these minor ticks blank (""):

```
minorx<-c(rep("",3))
axis(side=1, at=c(5,25,75), labels=minorx, tck=-0.02)
```

We use **rep** to create a list called '**minorx**' of three repeats of blank: "". We then provide this list as the labels for the minor ticks we add to the x-axis. We also make our minor ticks half the length we made the major ticks (**tck**).

Step 7: Now we need to add the y-axis, but first we will create a list with which to rename our major tick labels. As mentioned before, currently R has as default the format 1e, where 1e means '10 to the power'. We can make things look much neater if we use superscript, and we do this with the function **expression**. Here, we list the labels we want as our y-axis major ticks:

```
ymajor <-c(0.1,100,expression(10^10),expression(10^25),
expression(10^45))
```

You might recall that we used **expression** to *sub*script text contained in square brackets [ ] in the earlier example from section 8.2. This time, we do not need to combine its use with **paste** or any separating punctuation (see Scientific

Approach 8.1 for advice on constructing multicomponent expressions), as we are only producing a single superscripted element. Here, **expression** tells R to *super*script the value following on from the upward arrow symbol ^. See section 8.6 for further use of **expression**.

Step 8: We now use this list as the labels for our y-axis major ticks, remembering to specify the corresponding tick positions with **at**:

```
axis(side=2, at=c(0.1,100,10^10,10^25,10^45), labels=
ymajor,las=2)
```

Notice, we add this axis to **side** 2, and use **las=2** to rotate the y-axis labels to be perpendicular to the axis so that they are easier to read.

Step 9: As an extra trick, we can add a line running through all of the (already ordered) data points using the **lines** function we used in sections 6.4 and 7.3.4:

```
lines(x,y,lty=1,lwd=1,col="darkmagenta")
```

Figure 8.4c has our fictitious data plotted on appropriate logarithmic x- and y-axes, with customized and more easily readable tick mark labels. Although the default logarithmic plot R produced, Figure 8.4b, clearly presented the exponential growth of the data, it is always worth taking the time to tidy up automatic axes where the figure can be made even more accessible to viewers with a few simple tweaks.

Additionally, you should always consider whether presenting your data on logarithmic axes is necessary to effective interpretation. Logarithmic scales can sometimes lead to graphical misinterpretation, as the scales on the axes are less intuitive and can confuse viewers about the relationship between the presented variables (Menge et al. 2018). You should only use log-scaled axes if they help illuminate one or more key features in your data, such as meaningful variation across multiple orders of magnitude (Menge et al. 2018). See the Further Reading for links that will give you more details on logarithmic functions and when they are useful to use. See also the online resource 'Authors' attempts for

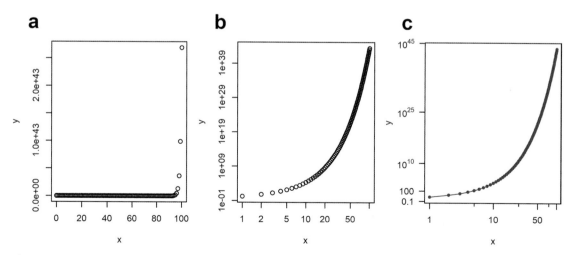

Figure 8.4 Fictitious exponential growth data, plotted: a) with linear x- and y-axes, b) with default logarithmic x- and y-axes, c) with customized logarithmic x- and y-axes and a line connecting all of the data points.

chapter 6' and associated 'R script for Authors' attempts chapter 6' for a good example of when to consider using log-scaled axes and when to log the data itself. If using log-scaled axes, you should consider the following as potentially handy tips to mitigate the possibility of misinterpretation:

- Presenting the same data on linear axes, as an additional panel in the figure or an inset (discussed in section 8.2) (Menge et al. 2018).
- Plot the 1:1 relationship line between the two variables, making it easier for viewers to assess the relative curvature and variation in the line of the data (Menge et al. 2018).

 Key point

If your data spans several orders of magnitude then the reader can appreciate variation in the low-valued data points more easily if you use a log axis.

8.6   Adding arrows, symbols, and shapes

Now, we're going to go over some additional features you can add to plots in R, sticking with our scatterplot example of imaginary exponentially increasing data from section 8.5. As with the inset bar chart example from sections 8.2–8.4, not all of these bits of code are specific to the one type of figure and you could use them for figures that are not scatterplots (although determining coordinates for positioning objects can involve more trial and error in figures with non-continuous axes).

Step 1: Starting with the end product of section 8.5 (Figure 8.4c), we will first draw an arrow within the plotting area. The template code for this is very simple: `arrows(x0,y0,x1,y1)`. The 0s indicate the coordinates where you want the end of the arrow to be and the 1s indicate the coordinates where you want the pointed front of the arrow to be. With the logarithmic scale, we have given ourselves more of a challenge in terms of figuring out where we want these coordinates to run, but the following values should provide a reasonable example:

```
arrows(2.5,10^9,3.14159,30, lwd=2,col="red",length=.1)
```

Our y-coordinates position the arrowhead at a fairly arbitrary value of 30 (close to but not overlapping the line through our data points) and the end of the arrow at 10^9 (using the values along the y-axis as a guide to keep the arrow fairly short in this case); you can see where this red arrow is positioned on Figure 8.5. Note that our second x-coordinate (where the pointed arrowhead is positioned) is a very specific value. This is not one of our data points; it relates to a particular mathematical constant that we will the label the arrow with next. `length` is a parameter for the size of the arrowhead.

You may have recognized the function **arrows** from the chapter 5 online resource 'Bar charts for quantitative multiple-samples data' and its accompanying 'R script' file. In these online materials, we used **arrows** to draw error bars onto a bar chart displaying continuous, quantitative data, by specifying flattened, error-bar-like arrowheads with the additional arguments **code=3** (which draws an arrowhead on each end) and **angle=90** (which draws caps as short lines perpendicular to the arrow shaft). Graphics can be almost endlessly tweaked to suit your purpose in R, so when you want to achieve something specific it is always worth investigating the additional arguments available to you in functions.

Step 2: Next, we will add a Greek letter 'π'–a character that is not easily accessible on keyboards to input into R code. Earlier, we used **expression** to present superscript (see section 8.5) and subscript (see section 8.2) in their proper format. Here we'll use **expression** alongside **paste** as part of a text command to add a label next to the arrow we just added to our scatterplot:

```
text(2.5,10^11,labels=expression(paste("x = ", pi)),
cex = 1.5)
```

You might remember our use of **paste** (without **expression**) in sections 2.2.1.2 and 2.2.2.1 when we presented values alongside % symbols as the labels for pie chart segments. In sections 6.3.3 and 7.3.5, we also used **paste** on its own to print out a calculated mean value with a neat label of 'mean=' on a scatterplot. The code here does essentially the same thing, but includes a mathematical symbol (making **expression** necessary) instead of a given or calculated value. It pastes 'x=' followed by the symbol π at the coordinates we specified (as you can see in Figure 8.5). R has a bank of Greek letters and other symbols that can be called by writing out their name in this way. See the Further Reading for a website listing other mathematical annotations available in R.

We here used **paste** as one method by which to join the separate parts of the text together (our 'x =' text and the Greek letter pi). You may recall that in section 8.2, instead of using **paste** and quotation marks, we used the space character ~ to separate out words of text within an **expression**. Either **paste** or various punctuation can be used to separate out multiple **expression** components, and which approach you choose will depend on whichever makes it easiest for you to carefully think through the presentation of the **expression** (see Scientific Approach 8.1 for a comparison of the two approaches). We recommend you check out the post on 'Expressions in R' recommended in the Further Reading if you want more explanation (and examples) of how to annotate figures with mathematical notation.

Step 3: Just as another example, let us now add another arrow, and this time a mathematical symbol rather than a Greek letter:

```
arrows(50,10^30,70,10^44, lwd=3,lty=2,col="forestgreen",
  length=.2)
text(45,10^28,labels=expression(infinity),cex=2.5)
```

This time our arrow is thicker, has a dashed line, and has a larger arrowhead (as you can see in Figure 8.5). Note that in our **text** neither **paste** nor any separating punctuation is needed in the **expression** as we are printing the symbol for infinity on its own.

Step 4: Just to demonstrate the complexity of superscripts and subscripts that can be coded for within an **expression** (with a little bit of trial and error), we're now going to add a completely irrelevant reaction equation, including a symbol to demonstrate that it's reversible! Try this:

```
text(5,10^40,labels=expression(paste("(Zn[OH]"[4],")"^{"2-"}
%<->% "ZnO + H"[2],"O + 2OH"^{"-"})),cex=1.2)
```

This code might look a bit daunting, but using **paste** it's really just a case of making sure your punctuation is in the right place. Superscripts need you to use a ^ symbol. If your superscript has a minus symbol, you'll need to contain the minus within curly brackets and quotation marks {"-"}. Subscripts need to be contained within square brackets [ ]. Every time you want to paste a superscript or subscript, you need to come out of the quotation marks "" that are telling R that the text should be printed at normal size. And the symbol we've used for the reversible reaction is made with the %<->% bit of code. This is more complicated than anything you are likely to need but, with patience, it can be done (take a look at Figure 8.5). Again, it is really worth checking out the link in the Further Reading if there's anything you want to try pasting onto R figures.

Step 5: Another feature you might want to add to your figures is a shape or shaded area. Sometimes with scatterplots, for example, the area represented underneath a line connecting the data points (or a line of best fit) is important to the graph's purpose, and so we may want to fill this area with colour. We can fill all the area under the line in our scatterplot by referring to the parameters of our x and y variables (here just 'x' and 'y') within the **polygon** function. The standard piece of code to do this, where you could swap out x and y for whatever your variable names are, would be:

```
polygon(c(min(x), x, max(x)), c(min(y), y, min(y))
```

However, recall that in section 8.5 our data contained a single x-value and a single y-value <= 0, which were both omitted from our logarithmic plot. Therefore, our minimum x- and y-values are not included in the figure. In their place, we will use the second value in each data list. To extract these values, we could subset each list to exclude values <=0, but instead we just ran 'x' and 'y' through the Console and copied and pasted the second value in each list to the necessary 'min' position in the **polygon** code:

```
polygon(c(1, x, max(x)), c(1.718282, y, 1.718282), col=
"chartreuse")
```

Step 6: It is also possible to shade only certain areas under the line. Here, we'll make the area under the line between x-axis values of 10 and 50 shaded a darker colour to demonstrate. To specify a shaded area, we first need to create lists of the values we want shaded:

```
xshade <- seq(10,50,1)
yshade <- y[c(seq(11,51,1))]
```

Our list **xshade** covers the range of x-axis values we want shaded, a sequence between 10 and 50. To specify the y-values that correspond with the x-axis values between 10 and 50, we use square brackets to extract from our list **y**. **y[1]** would be the first y-value in our y-data, so **y[3]** would be the third y-value and so on. We can use **c** and **seq** within the square brackets to create a list of positions from which to extract values from the **y** list. However, because the first value in our list of x-values was 0 (and omitted from the plot), the 11th value plotted is actually the one which will correspond to the 10 on our x-axis. Therefore, to get the corresponding values for our y-axis, we needed to extract the values from between positions 11 and 51 in our list **y** (which also started with a 0 that was omitted from the plot).

Step 7: We can now input these new lists into the **polygon** function in place of the previous 'x' and 'y'. Note, though, that the values in place of **min** for the y-axis specifications need to be kept as 1.718282 (the lowest plotted y-value) for our shading to reach the base of the plot:

```
polygon(c(min(xshade), xshade, max(xshade)), c(1.718282,
yshade, 1.718282), col="chartreuse4")
```

We can see in Figure 8.5 that both our shading polygons fit under our plotted line perfectly, thanks to our careful selection of values. We made things trickier by using data where values had been omitted from the plotting, but with most data the generic polygon code should work with less tinkering of values required.

Step 8: If we want to draw shapes using **polygon** that are unrelated to our data points, this involves simpler code. First, we need to specify the coordinates where we want the vertices of the polygon. It is important that you input the coordinates for the shape you want in order, either by following round the shape clockwise or anti-clockwise consistently, and that the order of the x-coordinate list corresponds with the order of the y-coordinate list—otherwise you can get some pretty crazy shapes! Here we'll add a random shape with seven vertices, so our x- and y-coordinate lists each have seven values:

```
shapex <- c(8,3,6,10,15,20,25)
shapey <- c(10^15,10^17,10^30,10^25,10^35,10^30,10^20)
```

Again, our exponential data and logarithmic axes mean that our y-coordinates appear more complicated than they would in most circumstances!

Step 9: Using these lists we can now plot our random shape with **polygon**:

```
polygon(shapex, shapey, col="pink",border="royalblue",lwd=4)
```

In Figure 8.5, you can see how we have customized this shape, filling it with the colour pink and surrounding it with a thick blue border. See the Further Reading if you are interested in seeing more examples for the application of **polygon** in R.

Step 10: As a final drawing trick, we will draw a couple of circles onto our figure. Circles in particular are one of the more common shapes added to figures, sometimes serving as a useful way to draw attention to a particular finding. Base R provides a way to draw circles through its **symbols** function, but with this function the size of circles is inconsistent across devices. Because we want consistent and accurate circles on our figure, we will instead use the **draw.circle** function available from the package 'plotrix' (Lemon 2006). With **draw.circle** we can specify the radii of circles using coordinates from a plot. So, first we need to install and activate **plotrix**:

```
install.packages("plotrix")
library(plotrix)
```

Step 11: Now, if our data was plotted on normal, linear axes, we could jump straight to the **draw.circle** function and plug in coordinates for a circle's position and radius. However, placing a circle onto a logarithmic plot is incredibly tricky, as the coordinates over which the radius spans are distorted. What we will do to counteract this, then, is to overlay our existing plot with a blank plot of known coordinates—we can then use the coordinates we specify for the plot to place our circle.

You will recognize features of the blank plot code below from section 8.4 and chapter 6, only now we set both x- and y-axes to run between 0 and 100 with intervals of 1:

```
par(new=TRUE)
plot(c(seq(0,100,1)),c(seq(0,100,1)), type="n", xaxt="n", yaxt=
  "n",xlab="", ylab="")
```

Step 12: The first circle we will plot will highlight an existing feature of our plot; that is, we will make its interior transparent so that it acts as a border alone, drawing attention to the object inside. To make shapes transparent, we will revisit the function **transparent** from the package 'yarrr' (Phillips 2017) that we used in section 5.4.2. So let's install (if you haven't already installed) and activate 'yarrr':

```
install.packages("yarrr")
library(yarrr)
```

Step 13: As a reminder, to use **transparency** we simply have to enter any co-lour as **orig.col**, then enter how transparent we want to make it (from 0 to 1, opaque to fully transparent) as **trans.val**. So, using the coordinates specified in the (essentially invisible) plot created above (and a bit of trial and error), we can now position a circle around the infinity symbol we plotted earlier:

```
draw.circle(82.5,63.5,radius=4,nv=1000,border="black",
  col=transparent(orig.col = "goldenrod1", trans.val = 1),
  lty=1,lwd=2)
```

The first two values of **draw.circle** are our x- and y-coordinates (based on our invisible 0–100 axes). We then give a value for the **radius**, and the number of vertices to draw the circle (**nv**)– 1000 gives a pretty smooth circle. You will see in Figure 8.5 that, although we specified the original colour of our circle's interior as 'goldenrod1', our transparency value of 1 makes the inside entirely transparent, so that we can read the underlying symbol.

Step 14: The second circle we will plot will be a filled circle that is unrelated to any of the existing features of the plot. For this shape, we continue to use our

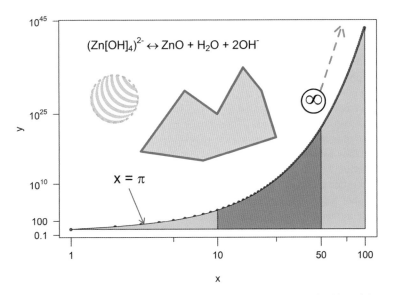

Figure 8.5 Fictitious exponential growth data, plotted on customized logarith-mic axes. With additional features: arrows, text including Greek letters and mathematical symbols, text including superscript and subscript, polygons for shading, a random polygon, a circle highlighting a feature, and a random circle.

invisible linear axes for the coordinates but no longer require a transparent interior. Instead, we will jazz up this circle using **density** and **angle** arguments:

```
draw.circle(15,65,radius=8,nv=1000,border="orange",col=
"seagreen1",lty=3,lwd=3,density=50,angle=45)
```

You might recognize **density** and **angle** from section 3.6. **density** refers to the density of shading lines, in lines per inch, while **angle** gives the angle of the pattern filling the shape. These arguments can also be included in the **polygon** function, but not when drawing on a logarithmic scale.

Our final Figure 8.5 is replete with fancy features and tricks that can be used when graphing in R, including: text with Greek letters, mathematical symbols, superscript and subscript, arrows, and shapes. If you wanted to add other shapes (such as squares, rectangles, or stars), you should look into R's **symbols** function further (see Further Reading), but the shapes we have covered should in themselves meet most of your graphing needs. As in section 8.4, we do not in any way advise that you use all possible customization tricks available to you in any single figure! We have done so here only to demonstrate what can be done. Which options are useful will depend on your data, your audience, and what you intend to communicate with your figure; remember the 'simple but not too simple' rule of thumb when designing your figures (discussed in section 1.7.1).

Key point

Annotating your graph can be an effective way to draw the reader's attention to features you think are of interest.

8.7   Annotating images and geographic maps

Sometimes, you may want to produce a figure that is simply a (often annotated) photograph of an experimental set-up or a map of a study site—without showing any data values at all. This is easy to do in R too using the techniques we have covered in this chapter. Specifically, the **par(fig)** code covered in section 8.4 can be used to position an image as a plot background, onto which a dummy plot with no data or axes can be added. Using the coordinates you set for the dummy plot, shapes (covered in section 8.6) and text can be added easily (see section 7.3.5 for details on adding text).

As a quick example, the following code produces Figure 8.6a, an annotated screenshot from an experiment using the PNG image file 'experiment' that we provide in the online resources:

```
library(png)
experiment <- readPNG(file.choose())
par(fig = c(0,1, 0, 1))
plot(c(seq(0,100,1)),c(seq(0,100,1)), type="n", xaxt="n",
  yaxt="n",xlab="", ylab="")
lim <- par()
rasterImage(experiment, lim$usr[1], lim$usr[3], lim$usr[2],
  lim$usr[4])
```

```
box(which = "plot", lty = 1)
arrows(78,12,55,38, lwd=2,col="skyblue",length=.1)
text(89,13,labels="Pea aphid",col="skyblue",font=2,cex=1)
text(89,6,labels=expression("("*italic(A.~pisum)*")"),
  col="skyblue",cex=0.8)
arrows(83,88,55,44, lwd=2,col="red",length=.1)
text(93,92,labels="2-spot\n Ladybird",col="red",font=2,cex=1)
text(93,82,labels=expression("("*italic(A.~bipunctata)*")"),
  col="red",cex=0.8)
segments(15, 52, 30, 99,col="goldenrod1",lwd=3)
segments(15, 52, 30, 1,col="goldenrod1",lwd=3)
segments(13, 52, 15,52,col="goldenrod1",lwd=3)
text(7,52,labels="Bean\nseedlings",col="goldenrod1",
  font=2,cex=1)
text(7,40,labels=expression("("*italic(V.~faba)*")"),
  col="goldenrod1",cex=0.8)
text(7,3,labels="Trial 13",col="white",font=2,cex=1)
```

Note: You might remember from sections 3.5.1 and 7.3.3 that including \n in text commands tells R to start a new line at this point in the text. The \n itself is not printed on the resulting figure, as you can see in Figure 8.6a. You will also notice our use of **expression**, the space character ~ and asterisks * in the text commands where we present species names in italics contained within non-italicized brackets—see Scientific Approach 8.1 for advice on the two main approaches for constructing multicomponent **expression** commands.

The following code produces Figure 8.6b, an annotated map of study sites using the PNG image file 'river_map' that we provide in the online resources:

```
library(png)
rivermap <- readPNG(file.choose())
par(fig = c(0,1, 0, 1))
plot(c(seq(0,100,1)),c(seq(0,100,1)), type="n", xaxt="n",
  yaxt="n",xlab="", ylab="")
lim <- par()
rasterImage(rivermap, lim$usr[1], lim$usr[3], lim$usr[2],
  lim$usr[4])
box(which = "plot", lty = 1)
segments(55, 85, 75, 85,col="goldenrod1",lwd=3)
segments(55, 85, 55, 82,col="goldenrod1",lwd=3)
segments(75, 85, 75, 82,col="goldenrod1",lwd=3)
text(65,93,labels="200m",col="goldenrod1",font=2,cex=1.75)
segments(12, 45, 17, 37,col="skyblue",lwd=5,lty=1)
segments(21, 85, 23, 74,col="skyblue",lwd=5,lty=1)
segments(58, 2, 62, 11,col="skyblue",lwd=5,lty=1)
text(20,47,labels="A",col="skyblue",font=2,cex=1.5)
text(27,87,labels="B",col="skyblue",font=2,cex=1.5)
text(55,11,labels="C",col="skyblue",font=2,cex=1.5)
library(plotrix)
draw.circle(8,49,radius=2,nv=1000,border="orange",col="red",
  lwd=1)
draw.circle(20,92,radius=1,nv=1000,
  border="orange",col="red",lwd=1)
```

```
draw.circle(66,17,radius=2.5,nv=1000,
  border="orange",col="red",lwd=1)
```

It is important to remember that absolutely anything is possible when graphing in R! There are so many resources available online, that with a bit of time and effort you can produce some truly impressive figures. We encourage you to check out some of the websites given in the Further Reading for inspiration.

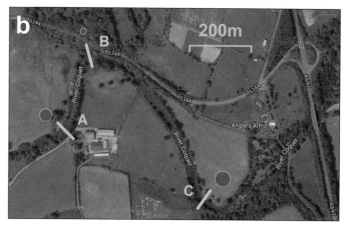

Figure 8.6 Images annotated using text and shapes in R: a) a screen-shot from an experiment investigating insect behaviour on plants, with the plants and insect individuals labelled, and the trial number stated in the bottom-left of the figure; b) a map (Google Maps 2020) showing three study sites (A–C) where hypothetical cross-sectional data was collected at meanders along the River Coquet, Northumberland. A scale bar has been added using the **segments** function (see section 7.3.4 for details), and circles indicate the relative mean size of pebbles collected at the deepest point of each site.

Key point

R can help make useful figures without data too, so can often help with your methods section as well as your results.

Chapter Summary

- As mentioned in chapter 7, R is a hugely flexible graphics package, with a vast number of features that can be modified when producing figures.
- By modifying plot space parameters, you can produce high-quality figures with inset plots; these can help you avoid misleading readers when using a non-zero y-axis.
- Design components, such as frames, images, background images, arrows, symbols, and shapes can be added to figures with relatively little effort.
- Logarithmic axes are an effective way to present data that spans several orders of magnitude.
- Images and geographic maps can be annotated in R by following the same methods used when annotating graphs.

Online Resources

The following online resources are available for this chapter at www.oup.com/he/humphreys-obp1e:

- R script for chapter 8
- canada_emissions.csv
- cow.png
- pig.png
- chicken.png
- Flag-Canada.jpeg
- experiment.png
- river_map.png

Further Reading

- 'Mathematical Annotation in R' provides a table of the syntax to use for a range of mathematical features, symbols, and operations: https://stat.ethz.ch/R-manual/R-devel/library/grDevices/html/plotmath.html
- 'Expressions in R' gives helpful examples of how **expression** and related functions in R can be used to annotate figures with mathematical notation: https://quantpalaeo.wordpress.com/2015/04/20/expressions-in-r/

- 'Axis Labels in R Plots Using expression() Command' explains how you can use features such as bold, italic, and superscript in your axis labels:
  https://www.dataanalytics.org.uk/axis-labels-in-r-plots-using-expression/
- 'Logarithmic Functions' presents a detailed look at logarithms in biological science, and how to plot data using logarithmic scales:
  http://www.biology.arizona.edu/biomath/tutorials/log/Log.html
- '5 Reasons to Love Logarithms' explains why logarithms can be very useful in biological science:
  http://ewanbirney.com/2013/08/5-reasons-to-love-logarithms.html
- 'Logarithms and Exponentials' gives a run-down of the R syntax for working with logs and exponentials:
  https://web.mit.edu/~r/current/arch/i386_linux26/lib/R/library/base/html/Log.html
- An interesting study into how logarithmic scales can be misinterpreted: D. N. L. Menge, A. C. MacPherson, T. A. Bytnerowicz, et al., Logarithmic scales in ecological data presentation may cause misinterpretation. *Nature Ecology and Evolution*, 2 (2018), 1393–1402.
  https://doi.org/10.1038/s41559-018-0610-7
- 'R polygon Function | 6 Example Codes (Frequency & Density Plot)'—uses examples to demonstrate the code to plot polygons, coloured borders of polygons, frequency and density polygons:
  https://statisticsglobe.com/r-polygon-function-plot/
- 'symbols: Draw Symbols (Circles, Squares, Stars, Thermometers, Boxplots)' explains more on how to use the `symbols` function to create and customize various shapes and features:
  https://www.rdocumentation.org/packages/graphics/versions/3.6.2/topics/symbols
- 'draw.circle: Draw A Circle' lists the code you need to draw a circle on an existing plot and customize its size and appearance using the `draw.circle` function:
  https://www.rdocumentation.org/packages/plotrix/versions/3.7-7/topics/draw.circle
- 'Quick-R: Advanced Graphs' by Datacamp includes links to a range of pages describing how to make the most of R's graphing capabilities by customizing your graphs, and looking at more statistically sophisticated graphs:
  https://www.statmethods.net/advgraphs/index.html
- 'The magick Package: Advanced Image-Processing in R' introduces how to install and use this toolkit for image processing in R, allowing you to tap into a comprehensive open-source library of different effects:
  https://cran.r-project.org/web/packages/magick/vignettes/intro.html
- A collection of around 400 charts created with R, 'The R Graph Gallery':
  https://www.r-graph-gallery.com/

GLOSSARY

argument (in R): optional or essential inputted parameters that a function requires to execute commands. Some functions contain many arguments, others contain no arguments. Arguments in R can have default values when not specified by a user.

caption: often equivalently referred to as 'figure legends', figure captions are concise text accompaniments to figures that provide readers with all of the information they require to understand a figure fully without recourse to the main text of the document containing the figure. See section 1.2 and Scientific Approach 1.2 for advice on writing effective captions.

categorical data: see its synonym: qualitative data.

command (in R): code run through the R Console; the line that code is typed into is called the command line.

continuous data: a type of quantitative data wherein the measurements that constitute the data could theoretically have an infinite number of possible values within a range depending on how fine-scaled your measurement equipment is, e.g. height. See section 1.3.1 and 'Data types' from the online resources for chapter 1 for more details.

correlation: a mutual relationship or connection between the values of two or more variables.

discrete data: a type of quantitative data wherein the measurements that constitute the data include only integers and the scale on which measurements were recorded constitutes a finite number of values that cannot be meaningfully subdivided, e.g. number of children in a household. See section 1.3.1 and 'Data types' from the online resources for chapter 1 for more details.

distribution: the frequency of all the different values of a variable in a data set, typically ordered from smallest to largest when displayed graphically.

function (in R): lines of code organized together to carry out a specific task. R has many built-in functions, but users can also define their own functions or use functions that others have made available online. Some functions require users to supply values or other information to their arguments in order to accomplish certain actions.

grid lines: horizontal and/or vertical lines that cross the plotting area to show the divisions of the axis or axes. Grid lines (sometimes called 'graph lines') come in two types, major and minor, much as you would find on graph paper. They can be used to help viewers interpret measurements from figures.

legend (in R): the `legend` function can be used in R to add a legend (aka a 'key') to a figure where it is needed to interpret the figure (e.g. to explain what different colours/points/line types represent). There are a few default arguments that users must supply the `legend` function with in order for it to successfully execute the command, but there are also many optional arguments that can be included for customization purposes (see section 7.4 for details).

list (in R): a list (or 'vector') in R can be created using the `c()` function. The letter c stands for 'concatenation', which essentially means 'bring together', and the function will bring together whatever elements are contained within the brackets (so long as they are of the same type, e.g. integer or character) to form an object that you can assign a name to (using the `<-` assignment character). See 'R Basics' from the online resources for chapter 1 for practice in creating lists.

multiple-samples data: data that comes from two or more related samples, with each sample from a different group of individuals, but for each individual we are still just interested in a single measurement.

nominal data: a type of qualitative data wherein the different categories to which the counts or observations are assigned have no logical rank order to them, e.g. blood group. See section 1.3.2 and 'Data types' from the online resources for chapter 1 for more details.

ordinal data: a type of qualitative data wherein the different categories to which the counts or observations are assigned have a logical rank ordering to them, e.g. shoe size. See section 1.3.2 and 'Data types' from the online resources for chapter 1 for more details.

outlier: a data point that seems unusual or noteworthy in the data set.

package (in R): free add-on collections of functions and data sets that can be installed and activated to improve existing base R statistical or graphical functionalities, or add new ones. See 'R Basics' from the online resources for chapter 1 for guidance on installing and activating packages.

parameters (in R): graphical parameters in R refer to the arguments and values that define or establish the positioning, sizing, and other features of plots. The function **par** (think 'par-ameters') can be used to edit R's vast range of customizable graphical parameters, including plot boundaries and organization of multiple plots/panels within a single figure. See section 7.2 for more on using **par**.

single-sample data: data that comes from one particular sample where for each individual within the sample there is one single measurement in the data set.

subset (in R): the **subset** function in R can be used to select and/or exclude particular variables and observations from a data set.

qualitative data: data made up of counts arranged into categories. Qualitative data can be ordinal or nominal. See section 1.3.2 and 'Data types' from the online resources for chapter 1 for more details.

quantitative data: data that is measured on a numerical scale. Quantitative data can be continuous or discrete. See section 1.3.1 and 'Data types' from the online resources for chapter 1 for more details.

trend: the general direction, or pattern, shown in data.

REFERENCES

Chapter 1

BATEMAN, S., MANDRYK, R. L., GUTWIN, C., GENEST, A. M., MCDINE, D., and BROOKS, C. 2010. Useful junk? The effects of visual embellishment on comprehension and memorability of charts. *Proceedings of the ACM Conference on Human Factors in Computing Systems (CHI 2010)*. Atlanta, GA, USA, 2573–2582.

FEW, S. 2011. The chartjunk debate: A close examination of recent findings. *Perceptual Edge Visual Business Intelligence Newsletter*, April, May, and June.

GARNIER, S., ROSS, N., RUDIS, R., CAMARGO, A. P., SCIANI, M., and SCHERER, C. 2021. Rvision–Colorblind-Friendly Color Maps for R. R package version 0.6.0.

OECD. 2020. *Gross Domestic Spending on R&D (Indicator)*, doi: 10.1787/d8b068b4-en. Available: https://data.oecd.org/rd/gross-domestic-spending-on-r-d.htm [Accessed 26/10/2020].

SU, Y.-S. 2008. It's easy to produce chartjunk using Microsoft® Excel 2007 but hard to make good graphs. *Computational Statistics & Data Analysis*, 52, 4594–4601.

WICKHAM, H. 2016. *ggplot2: Elegant Graphics for Data Analysis*. New York: Springer-Verlag.

Chapter 2

ADVISORY BOARD. 2012. *Which Workplace Surfaces Harbor the Most Germs?* Available: https://www.advisory.com/daily-briefing/2012/05/24/which-workplace-surfaces-harbor-the-most-germs [Accessed 26/10/2020].

BBC NEWS. 2017. *Seven Charts that Explain the Plastic Pollution Problem*. Available: https://www.bbc.co.uk/news/science-environment-42264788 [Accessed 26/10/2020].

BBC NEWS. 2020. *Climate Change: Where We Are in Seven Charts and What You Can Do to Help*. Available: https://www.bbc.co.uk/news/science-environment-46384067 [Accessed 26/10/2020].

GALLUP. 2017. *In US, Belief in Creationist View of Humans at New Low*. Available: https://news.gallup.com/poll/210956/belief-creationist-view-humans-new-low.aspx [Accessed 26/20/2020].

IPSOS MORI. 2018. *Public Attitudes to Animal Research in 2018*. London: Office for Life Sciences.

LI, J., LOVATT, M., EADIE, D., DOBBIE, F., MEIER, P., HOLMES, J., HASTINGS, G., and MACKINTOSH, A. M. 2017. Public attitudes towards alcohol control policies in Scotland and England: Results from a mixed-methods study. *Social Science & Medicine*, 177, 177–189.

YOUGOV. 2019. *Third of Brits Would Reintroduce Wolves and Lynxes to the UK, and a Quarter Want to Bring Back Bears*. Available: https://yougov.co.uk/topics/science/articles-reports/2020/01/28/third-brits-would-reintroduce-wolves-and-lynxes-uk [Accessed 26/10/2020].

Chapter 3

ALZHEIMER'S ASSOCIATION. 2020. Alzheimer's disease facts and figures. *Alzheimer's & Dementia*, 16, 391–460.

CDC WONDER. n.d. *CDC WONDER Online Database: About Underlying Cause of Death, 1999–2018*. US Department of Health and Human Services, Centers for Disease Control and Prevention, National Center for Health Statistics. Available: https://wonder.cdc.gov/ucd-icd10.html [Accessed 07/05/2020].

ENVIRONMENT AND CLIMATE CHANGE CANADA. 2020. *Greenhouse Gas Sources and Sinks: Executive Summary 2020*. Available: https://www.canada.ca/en/environment-climate-change/services/climate-change/greenhouse-gas-emissions/sources-sinks-executive-summary-2020.html [Accessed 10/07/2020; summary updated each year].

FORESTRY COMMISSION. 2003. *The National Inventory of Woodland and Trees–Great Britain*. Available: https://www.forestresearch.gov.uk/tools-and-resources/national-forest-inventory/national-inventory-of-woodland-and-trees/ [Accessed 10/07/2020].

HARRELL Jr, F. E. et al. 2021. *Hmisc: Harrell Miscellaneous*. R package version 4.5-0. Available: https://CRAN.R-project.org/package=Hmisc [Accessed 05/01/2022].

INARI, N., NAGAMITSU, T., KENTA, T., GOKA, K., and HIURA, T. 2005. Spatial and temporal pattern of introduced *Bombus terrestris* abundance in Hokkaido, Japan, and its potential impact on native bumblebees. *Population Ecology*, 47, 77–82.

TEJADA-VERA, B. 2013. *Mortality from Alzheimer's Disease in the United States: Data for*

2000 and 2010. Hyattsville, MD: National Center for Health Statistics.

Chapter 4

CRATSLEY, C. K. and LEWIS, S. M. 2003. Female preference for male courtship flashes in *Photinus ignitus* fireflies. *Behavioral Ecology*, 14, 135–140.

SHENG, Y., SOTO, J., ORLU GUL, M., CORTINA-BORJA, M., TULEU, C., and STANDING, J. F. 2016. New generalized poisson mixture model for bimodal count data with drug effect: An application to rodent brief-access taste aversion experiments. *CPT: Pharmacometrics & Systems Pharmacology*, 5, 427–436.

URBAN, M. C. 2015. Accelerating extinction risk from climate change. *Science*, 348, 571–573.

VENABLES, W. N. and RIPLEY, B. D. 2002. *Modern Applied Statistics with S.* New York: Springer.

WHITLOCK, M. C. and SCHLUTER, D. 2014. *The Analysis of Biological Data.* 2nd ed. New York: W. H. Freeman.

Chapter 5

AUDUBON. 2020. *Christmas Bird Count.* Available: https://netapp.audubon.org/CBCObservation/Historical/ResultsByCount.aspx# [Accessed 05/01/2022].

BEALL, C. M., DECKER, M. J., BRITTENHAM, G. M., KUSHNER, I., GEBREMEDHIN, A., and STROHL, K. P. 2002. An Ethiopian pattern of human adaptation to high-altitude hypoxia. *Proceedings of the National Academy of Sciences*, 99, 17215.

CHATFIELD, C. 1982. Teaching a course in applied statistics. *Journal of the Royal Statistical Society. Series C (Applied Statistics)*, 31, 272–289.

DRAPER, N. R. and SMITH, H. 1981. *Applied Regression Analysis.* New York: John Wiley & Sons.

PHILLIPS, N. 2017. *yarrr: A Companion to the e-Book 'YaRrr!: The Pirate's Guide to R'.* R package version 0.1.5. https://CRAN.R-project.org/package=yarrr [Accessed 05/01/2022].

SNEDECOR, G. W. 1956. *Statistical Methods.* Ames, IA: Iowa State College Press.

WILSDON, C. 2009. *Pigs.* Pleasantville, NY: Gareth Stevens Publishing LLLP.

Chapter 6

BOECKMANN, A. J., SHEINER, L. B., and BEAL, S. L. 1994. NONMEM Users Guide: Part V, NONMEM Project Group, University of California, San Francisco.

CORRELL, M., BERTINI, E., and FRANCONERI, S. 2020. Truncating the y-axis: Threat or menace? *Proceedings of the 2020 CHI Conference on Human Factors in Computing Systems.* Honolulu, HI: Association for Computing Machinery, 1–12.

DIGGLE, P. J. 1990. *Time Series: A Biostatistical Introduction.* Oxford: Oxford University Press.

HITE, N. J., GERMAIN, C., CAIN, B. W., SHELDON, M., PERALA, S. S. N., and SARKO, D. K. 2019. The better to eat you with: Bite force in the naked mole-rat (*Heterocephalus glaber*) is stronger than predicted based on body size. *Frontiers in Integrative Neuroscience*, 13.

NOAA NATIONAL CENTERS FOR ENVIRONMENTAL INFORMATION. 2020. *Climate at a Glance: Global Time Series.* Available: https://www.ncdc.noaa.gov/cag/global/time-series [Accessed 28/09/2020].

TRELOAR, M. A. 1974. Effects of puromycin on galactosyltransferase in Golgi membranes. MSc thesis, University of Toronto.

WISE CAMPAIGN. 2019. *2019 Workforce Statistics—One million women in STEM in the UK.* Available: https://www.wisecampaign.org.uk/statistics/2019-workforce-statistics-one-million-women-in-stem-in-the-uk/ [Accessed 28/09/2020].

WITTON, M. P. and HABIB, M. B. 2010. On the size and flight diversity of giant pterosaurs, the use of birds as pterosaur analogues and comments on pterosaur flightlessness. *PLOS ONE*, 5, e13982.

Chapter 7

HARRELL Jr, F. E. et al. 2021. *Hmisc: Harrell Miscellaneous.* R package version 4.5-0. Available: https://CRAN.R-project.org/package=Hmisc [Accessed 05/01/2022].

Chapter 8

ENVIRONMENT AND CLIMATE CHANGE CANADA. 2020. *Greenhouse Gas Sources and Sinks: Executive Summary 2020.* Available: https://www.canada.ca/en/environment-climate-change/services/climate-change/greenhouse-gas-emissions/sources-sinks-executive-summary-2020.html [Accessed 10/07/2020; summary updated each year].

GOOGLE MAPS. 2020. *Anglers Arms, Longframlington.* Available: https://www.google.co.uk/maps/@55.2819047,-1.7889789,9674m/data=!3m1!1e3 [Accessed 28/09/2020].

LEMON, J. 2006. Plotrix: A package in the red light district of R. *R-News*, 6, 8–12.

MENGE, D. N. L., MACPHERSON, A. C., BYTNEROWICZ, T. A., QUEBBEMAN, A. W., SCHWARTZ, N. B., TAYLOR, B. N., and WOLF, A. A. 2018. Logarithmic scales in ecological data presentation may cause misinterpretation. *Nature Ecology & Evolution*, 2, 1393–1402.

PHILLIPS, N. 2017. *yarrr: A Companion to the e-Book 'YaRrr!: The Pirate's Guide to R'*. R package version 0.1.5. Available: https://CRAN.R-project.org/package=yarrr [Accessed 05/01/2022].

STATISTA. 2020. *Per Capita Meat Consumption in Canada by Type 1998–2019*. Available: https://www.statista.com/statistics/442461/per-capita-meat-consumption-by-type-canada/#:~:text=Chicken%20was%20the%20most%20consumed,-pounds%20per%20capita%20since%201998 [Accessed 12/10/2020].

URBANEK, S. 2013. *png: Read and write PNG images*. R package version 0.1–7. https://CRAN.R-project.org/package=png [Accessed 05/01/2022].

URBANEK, S. 2019. *jpeg: Read and write JPEG images*. R package version 0.1–8.1. https://CRAN.R-project.org/package=jpeg [Accessed 05/01/2022].

INDEX